JOURNAL OF CYBER SECURITY AND MOBILITY

Volume 3, No. 3 (July 2014)

Special Issue on
Big Data Theory and Practice

Guest Editors:
Jiang Bian
Remzi Seker
Kenji Yoshigoe
Mengjun Xie

JOURNAL OF CYBER SECURITY AND MOBILITY

Aim
Journal of Cyber Security and Mobility provides an in-depth and holistic view of security and solutions from practical to theoretical aspects. It covers topics that are equally valuable for practitioners as well as those new in the field.

Scope
The journal covers security issues in cyber space and solutions thereof. As cyber space has moved towards the wireless/mobile world, issues in wireless/mobile communications will also be published. The publication will take a holistic view. Some example topics are: security in mobile networks, security and mobility optimization, cyber security, cloud security, Internet of Things (IoT) and machine-to-machine technologies.

Published, sold and distributed by:
River Publishers
Niels Jernes Vej 10
9220 Aalborg Ø
Denmark

Tel.: +45369953197
www.riverpublishers.com

Journal of Cyber Security and Mobility is published four times a year.
Publication programme, 2014: Volume 3 (4 issues)

ISSN 2245-1439 (Print Version)
ISSN 2245-4578 (Online Version)
ISBN 978-87-93237-21-6 (this issue)

JOURNAL OF CYBER SECURITY AND MOBILITY COMMUNICATIONS

Volume 3, No. 3 (July 2014)

Foreword by Guest Editors

We are now in the era of the Big Data revolution where nearly every aspect of computing engineering is driven by increasingly large, complex, and diverse datasets. Big Data presents not only a world of new opportunities but also new challenges. With threats multiplying exponentially, the ability to gather and analyze massive information will be a decisive factor in the battle against malicious software and adversaries. It is natural to consider a cloud-computing environment to address the computational requirements for big data analytic applications. However, it is equally important to address the security concerns in terms of policies, technologies, and controls deployed to protect data, applications, and the associated infrastructure of cloud computing. In this special issue of JCSM on Big Data, the article by Miller et al. focuses on technical and policy infrastructure for digital forensic analysis in the cloud as cyber-crime is a growing trend around the world. Further, the massive collections of imagery on the Internet have inspired a stream of interesting work on image processing related big data topics. The article by Shen et al. describes their novel structure-based image completion algorithm for object removal while maintaining visually plausible content with consistent structure and scene texture in photos. Such novel technique can benefit a diverse range of applications, from image restoration, to privacy protection, to photo localization. A surge of graph-computing frameworks has appeared in both academia and industry to address the needs of processing complex and large graph-structured datasets, where each has its respective benefits and drawbacks. Leveraging the right platform for the right task is daunting for users of these frameworks. A review by Zhao et al. provides the context for selecting the right graph-parallel processing framework given the tasks in hand. They have studied several popular distributed graph-computing systems aiming to reveal the characteristics of those systems in performing common graph algorithms with real-world datasets. Their findings are extremely informative.

Jiang Bian
Division of Biomedical Informatics
Univerisity of Arkansas for Medical Sciences

jbian@uams.edu

Remzi Seker
Department of Electrical, Computer, Software, and Systems Engineering
Embry-Riddle Aeronautical University - Daytona Beach Campus
sekerr@erau.edu

Kenji Yoshigoe
Department of Computer Sciences
University of Arkansas at Little Rock
kxyoshigoe@ualr.edu

Mengjun Xie
Department of Computer Sciences
University of Arkansas at Little Rock
mxxie@ualr.edu

Foreword by Editor-in-Chief

It is great pleasure to introduce you to the July 2014 issue of Journal of Cyber Security and Mobility. This is issue number 3 and volume 3 of the series. This issue consists of five papers including three papers from special issue on "Big Data Theory and Practice" and two papers from open call. Guest editors of the special issue Prof. Jiang Bian, Prof. Remzi Seker, Prof. Kenji Yoshigoe, Prof. Mengjun Xie have provided the details of the three papers that are part of the special issue as part of guest editors' foreword separately. Those three papers highlight the importance of digital forensic analysis in the cloud, novel structure-based image completion algorithm and distributed graph computing systems, respectively. Readers are encouraged to read those papers to understand the issues and development around Big Data. The fourth paper in the journal is the first paper from the open call. In this paper, titled, "A Cached Registration Scheme for IP Multimedia Subsystem (IMS)," Al-Doski and Mohan propose and analyze a cached registration scheme to reduce the delay associated with registration in IMS (IP Multimedia Subsystem). In addition, this paper also studies the impact of user movement patterns on IMS components in terms of user registration. Overall goal of this paper is to discuss novel mechanisms to increase the quality of end user experience in a highly mobile world. In the fifth paper, titled," Personal Denial of Service Attacks (PDOS) and Online Misbehavior: The Need for Cyber Ethics and Information Security Education on University Campuses," Podhradsky et al. underscore the need to provide basic information security and cyber ethics training for all university students to address a new type of cyber crime called, Personal Denial of Service Attack (PDOS) where an individual deliberately prevents the access of another individual or small group to online services. While there are a number of cybersecurity programs in US at a corporate level, the authors also highlight the importance of more qualified graduates with cybersecurity expertise and cyber ethics that can only be introduced in the university. We would like to thank the guest editors for bringing out articles on such as important area of Big Data and the authors of all the articles to raise the awareness and discuss the issues in the area of cybersecurity and mobility. We would like to thank the members of the steering board, advisors, reviewers of the articles and colleagues from River Publishers for their efforts towards the production of this issue. We hope the readers will enjoy these articles and we solicit contributions and guest editors for future issues of the journal.

Editors-in-Chief
Ashutosh Dutta, AT&T, USA
Ruby Lee, Princeton University, USA
Neeli R. Prasad, CTIF-USA and SAI Technology

Forensicloud: An Architecture for Digital Forensic Analysis in the Cloud

Cody Miller, Dae Glendowne, David Dampier and Kendall Blaylock

Distributed Analytics and Security Institute, Mississippi State University, Mississippi State, MS, USA
{miller; dae; dampier}@dasi.msstate.edu, kblaylock@cse.msstate.edu

Received 30 June 2014; Accepted 20 August 2014
Publication 7 October 2014

Abstract

The amount of data that must be processed in current digital forensic examinations continues to rise. Both the volume and diversity of data are obstacles to the timely completion of forensic investigations. Additionally, some law enforcement agencies do not have the resources to handle cases of even moderate size. To address these issues we have developed an architecture for a cloud-based distributed processing platform we have named Forensicloud. This architecture is designed to reduce the time taken to process digital evidence by leveraging the power of a high performance computing platform and by adapting existing tools to operate within this environment. Forensicloud's Software and Infrastructure as a Service service models allow investigators to use remote virtual environments for investigating digital evidence. These environments allow investigators the ability to use licensed and unlicensed tools that they may not have had access to before and allows some of these tools to be run on computing clusters.

Keywords: digital forensics, parallelization, cloud computing, cloud forensics, virtualization, virtual desktop infrastructure, HPC, cluster, infrastructure as a service, software as a service.

Journal of Cyber Security, Vol. 3, 231–262.
doi: 10.13052/jcsm2245-1439.331

1 Introduction

Cyber-crime is a growing trend in the U.S. and around the world. More cyber-crimes are being committed every day, from e-bay fraud to cyber-extortion [27]. Additionally, the amount of data that must be processed in a digital forensic examination continues to rise at a very high rate. Part of this surge is due to the increased storage capacity of hard disk drives. A typical personal computer will often contain a 1 TiB drive with options for expanding to 2, 3, or even 4 TiB in some cases. Computers employed in commercial or government organizations can have even more drives. Add the fact that an investigation may encompass multiple systems, include the possibility of network, live response, and memory data; the processing time for a digital forensics examination rapidly becomes overwhelming. Considering that the rate of computer adoption is not slowing down, this trend appears to be continuing well into the future.

Over the years there has been a significant amount of research calling for increased processing power being applied to digital forensics [29] [31] [32] as well as an improvement of the current tools and techniques we are applying [19] [20] [26]. Research has shown that processing time can be decreased via both methods, increased processing power and more intelligent techniques, and they should be employed together for maximum effectiveness. This research indicates that without a way to take advantage of more processing power and improved tools, it is unlikely that the digital forensics community will be able to keep up with the demand for its services. Another aspect of digital forensics is that centralized processing laboratories at state and federal levels are not suited to taking on an ever-increasing number of local criminal cases containing digital evidence. The backlog at these centralized labs is already long, and increasing the workload will not help to decrease the backlog. Low-priority cases, such as those not involving child victims or significant losses, are probably never going to make it to the top of the queue, as new higher priority cases will get processed first.

This paper presents an architecture for a cloud-based digital forensics processing platform named Forensicloud. It explores the issues, both technical and judicial, related to performing digital forensics in a remote environment. Finally, it presents a test plan for evaluating various components of a Forensicloud implementation. The focus of this paper is using the cloud to perform digital forensics not performing digital forensics of the cloud.

1.1 Contributions

In this paper, we describe the architecture for a cloud-based digital forensics analysis platform. Additionally, we identify challenges, both technical and judicial, regarding the implementation of a cloud-based digital forensics analysis platform. We identify a set of tests that can be performed to evaluate various components of a Forensicloud. The platform takes advantage of a distributed-computing environment to provide faster processing capability for performing digital forensics investigations remotely. We believe this will benefit small law enforcement organizations that could not otherwise afford to purchase their own comparable computing resources to perform in house investigations.

1.2 Motivations

This effort is motivated by over five years of experience building department sized digital forensics laboratories in rural Mississippi. During that time, the National Forensics Training Center (NFTC) at Mississippi State University was engaged in building small laboratories to provide digital forensics capabilities in strategic areas of Mississippi, where it was difficult for departments to either afford a lab or to engage the digital forensics lab in the Cyber-crime Fusion Center in Jackson, due to priorities of the central lab. In many cases, there were trained officers in these small departments, but without a lab of their own, they could not work their own cases. The efforts of the NFTC were to provide regional labs at larger departments where the smaller departments could time share on the equipment and use the larger department's lab to work their own cases. What was discovered is that when the larger department received the lab equipment, they used it a majority of the time, thereby making it difficult to provide the smaller departments with the time they needed in the lab.

Forensicloud was envisioned to allow any small department to get an account on the cloud server, and with only a small client at the local station, to work on their own cases using a greater level of computing resources. We acknowledge that a significant challenge is the upload of the media containing the potential evidence to the central server, where processing can be accomplished centrally. With normal Internet access, the time required to upload an average sized hard drive image is daunting. We have yet to come up with a long term solution to this problem, but a reasonable stop-gap measure is to have the department deliver the media to the location of the server for upload, or to a facility in the state with a high speed connection to the Internet for upload.

A Forensicloud environment will provide the law enforcement community substantial processing resources which could not be achieved by a stand-alone workstation environment. Even with the distribution of workstations throughout the state of Mississippi by the National Forensics Training Center, there is still, and will likely always be, an imbalance of utilization between different law enforcement agencies. The on-demand nature of a cloud environment is more robust to the varied processing needs of different law enforcement agencies. It will be able to accommodate and aid a range of departments from those that handle minimal digital forensics cases to those that see sudden surges in crimes requiring digital forensic processing. This versatility is a cornerstone of the motivation for developing Forensicloud.

The processing flexibility of a Forensicloud environment will range from being able to process a large number of cases at one time, therefore reducing backlog, to processing a low number of cases with greater processing power, therefore reducing the time of evidence processing. This distribution of resources can also be utilized for high profile cases, such as a child abduction, that must be processed quickly to provide investigators with time sensitive information that may be vital to the outcome of the situation. A Forensicloud environment will also provide a collaborative capability between departments and examiners that would otherwise be difficult in a stand-alone digital forensic environment.

2 Related Works

2.1 Digital Forensics using Cloud Environments

In [21] the authors identify some issues associated with hosting a digital forensics server in a cloud environment. However, most of the paper is about doing digital forensics of the cloud and not using a cloud-based analysis platform to perform digital forensics. They do propose the following issues that need to be addressed when using the cloud to perform digital forensics:

- the evidence should not change when transmitted to and from the cloud and should not change while stored
- local laws should be observed when storing evidence in the cloud
- unauthorized access (either physically or digitally) to the evidence should be prohibited and no one should be able to change the evidence
- only users that have authorization to the evidence should be able to access it

These are all valid points concerning the transmission, authentication, and handling of evidence during an investigation, but the authors do not offer any possible solutions to these issues.

In [33] the authors discuss a model called MapReduce that processes data across many clusters. They explain that the current tools process linearly not in parallel. They implement MPI MapReduce which uses Message Passing Interface (MPI) and Phoenix (a shared memory version of MapReduce) to make the current implementations of MapReduce more efficient. They tested their MPI version of MapReduce and showed that it outperformed Hadoop for CPU, memory, and I/O tasks.

2.2 Increasing Processing Power and Distributed Architectures

In [32], Roussev and Richard note the limits of traditional, single workstation, digital forensics tools and define a set of system requirements for distributed forensics tools. They use the following factors to show that distributed forensics tools are now necessary:

- storage devices are growing in capacity
- the I/O speed of these devices is growing slower than the capacity increases
- digital forensic tools are becoming more and more complex

They propose the following requirements for distributed digital forensics toolkit:

- scalability
- platform-independence
- lightweight
- interactive
- extensible
- robust

The authors made a distributed architecture for digital forensics using these requirements that showed live search improvements that were 18 times faster than a traditional workstation.

In [31] the authors argue that current forensic tools are insufficient because users have not specified performance requirements and that the developers of tools fail to make performance a priority. They suggest that real-time forensics and triage may be a solution. Real-time forensics and triage places a time limit on the computation. They suggest the following goals of acquisition and processing to achieve best performance:

- they should complete at approximately the same time
- their results should be presented immediately upon completion

The authors explain that file system metadata, block forensics, and similarity digests are extremely fast and that file attributes and Windows registry can be processed in under one hour. The authors state that because data carving generates more data to process and has a high false positive rate, it may be time to reconsider why we use carving and if we can achieve its most important results (recovery of files) by other means. With testing they noticed that only a few forensic processes could actually be done in real-time on a traditional computer (8-core). Using a server (48-core) more processes can be completed in the specified time limit. They also note that SSDs may require more resources, as the read speed is much faster than the HDDs used in their testing. They estimate that all processing (at 120 MiB/s read speed) could be completed within the one-hour time limit with two to four 48 or 64 core servers.

In [29] the authors improve a previous tool known as sdhash by using parallelization. They also demonstrate that the imaging phase of investigation can also be a processing phase by running their tool as the evidence is mirrored. Using their new tool (sdhash-dd), data can be represented at 1.6% of the original size. They showed that they can lookup a small file (16KB) at 1.4TiB/s and they have almost perfect true/false positive rate.

Foremost was turned into a parallelized program by use of a parallel API in [26] to improve the speed at which it processed data. Foremost searches for known file headers and footers on a disk image; it does this sequentially in which it retrieves a chunk of data from the image, searches the chunk, then retrieves another chunk. The authors create an API enabling open source programs to be parallelized. The API uses a communication arbiter that allows the API accesses to the disk image; it has several features such as data safeguards, caching, and data read-ahead. The API also uses a channelized task scheduler that schedules several subtasks to do work. While the parallelized Foremost used more memory than the serial version, it does increase execution speed by 2.5 times on average.

2.3 Better Algorithms for Digital Forensics

Bulk Extractor [20] operates on disks images, files, and directories, and memory dumps and extracts various types of digital evidence including IP addresses, credit card numbers, or user-defined regular expressions. The tool supports parallel execution to improve processing time. It reads the input from start to end and passes the data to scanners that identify the data.

Compressed data is decompressed and sent back through the scanners. Bulk Extractor then creates report files that contain the locations of the identified files on the input. The authors provide a GUI that creates a histogram of the report files allowing quick analysis. Bulk Extractor performed 10 times faster than EnCase when extracting email addresses.

In [19] the author explains that the current age of digital forensics is coming to an end and we need to go in a new direction for digital forensics research. The old ways of performing digital forensics and the tools used need to be updated. The author explains several challenges of current digital forensics research and proposes a new direction for digital forensics research. This new direction requires new data abstractions, modularization and composability of tools, new framework supporting alternate processing models, and support for scaling and validation.

2.4 Scalable Digital Forensics Frameworks

The Sleuth Kit Hadoop [12] is a framework that uses The Sleuth Kit (TSK) on top of Apache Hadoop. It has three phases that it uses to analyze data: ingest, analysis, and reporting. Ingest retrieves information about the file system and the files on the image. The analysis phase uses various modules of TSK to analyze the data. Finally, the reporting phase generates reports on the analysis. TSK Hadoop uses Apache Hadoop [17] to distribute the process of analysis across several nodes. Hadoop has an intergraded distributed file system, a job scheduler, and a Java implementation of MapReduce for parallel processing [17]. Using Hadoop and TSK together benefit from increased processing power from parallelization. However, using TSK Hadoop limits the number of tools to those supported by TSK. One of the goals of Forensicloud is to enable a variety of tools and techniques to function in the environment.

2.5 Virtualization Architectures

Several virtualization solutions were considered when designing Forensicloud. Most solutions did not meet all of the requirements needed by Forensicloud. Xen [16], KVM [7], and OpenVZ [11] have no native support for Virtual Desktop Infrastructure. OpenStack [10] and VMware ESXi [14] are the top two choices we have looked at that meet all the requirements natively. Microsoft Hyper-V [8] and Citrix XenServer [3] were also considered, however they were not as user-friendly, their installations were difficult, or their management software did not have as many features as VMware ESXi and OpenStack.

3 Requirements

There are several challenges that must be addressed for an effective implementation of Forensicloud. These challenges involve both technical capability and strict digital forensics processes. Each of the challenges listed below can be overcome while upholding the integrity of digital evidence and providing the user with a high level of digital evidence processing capability.

The Scientific Working Group on Digital Evidence Model Quality Assurance Manual for Digital Evidence Laboratories specifies rules that a digital forensics laboratory must follow when computers or automated equipment are used for the acquisition, processing, recording, reporting, storage or retrieval of examination data [25]. The rules are:

1. "Digital forensic tools are documented in sufficient detail and are suitably validated"
2. "The integrity and confidentiality of data entry, data storage, data transmission, and data processing is protected"
3. "Computers and automated equipment are maintained to ensure proper functioning and are provided with environmental and operating conditions necessary to maintain the integrity of examination data"
4. "Unauthorized access is prevented for computer systems used for examining digital evidence"

Each of these is a challenge that all digital forensic laboratories face, and Forensicloud is no different. Forensicloud satisfies these rules in the following ways:

1. Forensicloud will provide all documentation that each tool has available however tool validation will not be done as it is up to the investigator to determine the validity of a tool.
2. As described in the security measures below, the integrity of the data and prevention of unauthorized access will be possible.
3. All hardware and software utilized by a Forensicloud will be maintained and kept at a high level of operation.
4. Only users whom have access to Forensicloud will be able to access it and users will not be able to access data that is not apart of their case.

3.1 Client Security

One of the primary focuses of Forensicloud is the widespread availability of significant computing resources. This capability will allow officers to utilize hardware and software that would otherwise be difficult or too expensive for them to acquire and maintain. To provide this functionality it is imperative that officers be able to access Forensicloud to not only set up their desired evidence processing, but to also analyze the output from that processing. Access Forensicloud will be done over the Internet and must therefore be secured. To help ensure the security of the clients that are accessing Forensicloud will use traditional user access, such as confirmed users who will have unique login credentials. The client security will be taken a step further by requiring a client machine be approved and validated to access Forensicloud. These client machines will be limited to stand-alone machines that are in secure areas of a law enforcement or government office. These client machines must also be under the control of a trusted user that has been approved to use Forensicloud system and perform digital forensic investigations. Evidence and case data will only be accessible by the machine(s) and user(s) that are authorized to examine that specific case.

3.2 Data Security

As with any digital evidence investigation, the security of the evidence throughout the entire digital forensic process is paramount. The idea of Forensicloud naturally includes the transmission and remote storage of sensitive information. Maintaining the confidentiality and integrity of the data while ensuring it is available to forensics examiners is crucial. In order to protect the evidence being transmitted across the Internet data encryption will be used. The data will not be encrypted is when it is being processed, when the investigator is viewing it, when it is stored at Forensicloud and the local image on the investigator's computer. Since the processing engine is not open to the Internet and the evidence is isolated from other users, the data will be secure.

3.3 Network Latency

Unlike traditional digital forensic practices where the working copy of the evidence can be loaded on a stand-alone workstation in the examiners lab, Forensicloud will need to have the capability to receive digital evidence from remote locations. Currently, the average upload speed in the U.S. is 7.7 Mbps [6]. At this rate, it would take approximately 25 days to transmit 2 TiB of data across the Internet. With the advent of Internet2 speeds are significantly increased, which will allow for much faster upload capability. With the speed of Internet2 it will be possible for examiners to upload the working copy of their evidence to Forensicloud in a practical amount of time. Currently, there are 7 Internet2 participants distributed throughout the state of

Table 1 Theoretical transfer of files

Size of evidence	100 Mbit/s	10 Gbit/s
100 GiB	02:57:29	00:01:44
500 GiB	14:47:28	00:08:40
1 TiB	30:17:31	00:17:44
5 TiB	151:27:39	01:28:44

Mississippi [9]. Each of these locations has a 10 Gbps connection. Forensicloud will utilize these locations by having upload stations at each that have a 10 Gbps connection to the Forensicloud storage. Table 1 shows the theoretical time it would take to transfer different sizes of evidence with a 30% overhead on the network. These times do not represent reading from a disk and transmitting it across the Internet, they assume that there is no disk bottleneck. The actual transmission speed will be limited by the disk speed when using Internet2. Uploading the evidence is a one-time cost because the evidence will only be transmitted once and will be stored during the entire investigation of the case. The investigator will only download the reports generated by the tools used and will not download all the evidence.

The drawback of Internet2 is that fact that it is not available in every location and it may be unreasonable for some examiners to travel to an Internet2 location to upload their evidence. For a scenario where an examiner is unable to travel to an Internet2 or Forensicloud upload site location they will need to ship the working copy of their evidence to either an Internet2 location or directly to Forensicloud. This solution will work as long as the shipping provider will uphold chain of custody for the evidence.

3.4 Data Authentication

An important part of any digital forensics examination is the authenticity of the digital evidence and case files. This is no different in the Forensicloud environment. Cryptographic hashes, such as MD5, SHA1 and SHA256 will be used to authenticate the data. This will be done when the evidence files are sent to Forensicloud to ensure the data that is submitted is the same as the data that was originally seized by the local investigators. These hashes will also be used to ensure the data is not changed during the processing and storage of the evidence.

3.5 Data Storage

The purpose of the cloud is to leverage the greater resources of a high performance computing system while providing a plethora of tools to examiners. Based on the practices outlined in the Scientific Working Group on Digital

Evidence Model Quality Assurance Manual for Digital Evidence Laboratories Forensicloud will be a short-term data storage area and considered to be a working area for the digital investigation. Because of this, the evidence and case data that is stored on the cloud will only be stored while the case is considered to be an active investigation. Evidence and case files will not be retained longer than 90, unless special permission is given or further analysis is needed [25]. This permission will be given for special circumstances only and the approved user of the case must submit a request.

3.6 User Interface

The method in which the examiner interacts with Forensicloud is critical. It is imperative that the interface for the Forensicloud be both versatile and user-friendly. There are essentially two points of interaction between the investigator and a Forensicloud system. There will be an upload-and-request interface and a processing-and-review interface.

The upload-and-request interface will be a client that will run on the examiner's workstation and the upload facility workstation. This client will give the examiner the ability to create a new case, upload evidence files, and request specific processing items that he or she would like to have done to each evidence item.

The processing-and-review interface will give the examiner the ability to review the evidence as well as any results from processing. They will also have the capability to perform any additional processing that may be beneficial for their case. This interface will utilize a virtual machine to give the examiner the look and feel of a traditional digital forensic workstation. As a part of this look and feel, AccessData's FTK [5] and Guidance Software's Encase [4] will be available to the examiner. The availability of these two tools is important because they are two of the most widely used commercial tools for digital forensic examination [23]. Including both FTK and Encase is an advantage because it brings two of the most well known digital forensic tools to users that may not have access to those tools, either for economic reasons or technical capability. In addition to the tools that will be built into Forensicloud for processing the processing-and-review interface will allow the examiner to install their own tools into the virtual machine. This provides the examiner with the versatility that they would have with a stand-alone digital forensics workstation while still leveraging the greater computing power of a Forensicloud system.

Due to the fact that the majority of examiners use a graphical user interface (GUI) instead of command line tools [23], it is important that the upload-and-request interface and processing-and-review interface primarily utilize a GUI but have the flexibility and control for the examiner to use command line tools when needed. Both interfaces will be GUIs that provide the user with the options they will need for the tasks that are built into Forensicloud service. The virtual machine environment will allow the user to use tools that they are familiar with and interface, command line or GUI, that they are comfortable with.

Another benefit of Forensicloud is the ability for streamlined collaboration. The cases are stored in a centralized location, other examiners or investigators will have the ability to view or process the same case, as long as each are approved by the case manager. This allows for new examiners to get assistance from an experienced examiner without the need for one to travel to the other. This will also help with collaboration between departments when evidence items may be involved in multiple crimes.

3.7 Chain-of-Custody

As with any investigation, tracking digital evidence is crucial. The procedures for the use of Forensicloud will require that any evidence files that are uploaded must have chain of custody documentation. The same is true if an examiner or investigator makes a request to obtain the evidence files from the Forensicloud storage. In addition to chain-of-custody documentation, the system will maintain logs of all evidence and user events, such as logins, tool processing, and evidence upload and downloaded. These logs will automatically be generated and will contain a timestamp, a user id, and a report of the action.

4 Forensicloud Architecture

Cloud computing, as defined by NIST, is a model for enabling ubiquitous, convenient, on-demand network access to a shared pool of configurable computing resources (e.g., networks, servers, storage, applications, and services) that can be rapidly provisioned and released with minimal management effort or service provider interaction. This cloud model is composed of five essential characteristics, three service models, and four deployment models [24]. Essential Characteristics:

- On-demand self-service
- Broad network access

- Resource pooling
- Rapid elasticity
- Measured service

Service Models:

- Software as a Service (SaaS)
- Platform as a Service (PaaS)
- Infrastructure as a Service (IaaS)

Deployment Models:

- Private cloud
- Community cloud
- Public cloud
- Hybrid cloud

The architecture of Forensicloud will fulfill the five essential characteristics. Forensicloud will

- be accessible by investigators to analyze evidence as needed (on-demand self-service)
- be available to an entire state (broad network access)
- use virtualization to pool several servers resources (resource pooling)
- be able to expand or shrink based on the needs of its users (rapid elasticity)
- schedule processing based on priority and consumption of the investigator (measured service)

By providing software and an environment to use the provided software, Forensicloud uses both SaaS and IaaS models. The deployment model used will be a community cloud. In this instance, the community is the collection of law enforcement agencies within a given state. Digital forensics evidence by nature is sensitive information. As such, public clouds like Amazon's EC2 should not be used for either the investigation or processing components.

4.1 Architecture Overview

Forensicloud will enable an investigator to upload, process, and analyze evidence. Investigators will have access to a client that will allow them to connect to Forensicloud and upload digital evidence. The client will first authenticate the investigator then it will create a secure connection to Forensicloud. Using this connection, an investigator will set up a job by selecting the digital forensic tools that will be run. The investigator will then upload a disk image to Forensicloud using the client. While the image is being uploaded some of the selected forensic tools will run; these tools do not require the full disk image. Tools that require the entire image will be executed when the image has finish uploading. The tools will run on a cluster and the results will be saved to the job directory.

During the processing and analysis of the evidence, the investigator will authenticate and connect to a virtual machine remotely. The connection to the virtual machine will be secured. The investigator will only be able to connect to the virtual machine that has the evidence that they uploaded. As tools finish on the cluster their results become available in the virtual machine. In this virtual machine they have access to tools that cannot or do not benefit from being run on a cluster. The investigator can use the output of these tools as well as other tools on the virtual machine to analyze and investigate their case. Virtual machines are isolated from each other and the virtual machine is not able to access evidence that is owned by another investigator.

The analysis virtual machine and the cluster have access to a data store that has the evidence uploaded by the client. The evidence is encrypted from the client, decrypted on the cluster, and the results of the processing will be encrypted back to the data store. The analysis virtual machine will then decrypt the data for the investigator.

There are three main parts to Forensicloud: the client, the investigation component, and the processing component. The client will do initial job setup and will upload the evidence to Forensicloud. The investigation component provides the user with a virtual machine to use to analyze the evidence. The processing component will run forensics tools using a cluster.

4.2 Forensicloud Client

The client used to upload and setup jobs will be installed on the investigators computer. Once installed it will connect to Forensicloud and validate the system and itself. Validation requires the law enforcement agency to register their system(s) with the providers of Forensicloud. The client will also validate itself to ensure it is updated and has not been altered. Only departments that have registered will be capable of accessing Forensicloud. The investigator must also authenticate with Forensicloud. If the client is valid and the investigator is authenticated a secure connection will be made and the upload and job setup can continue. The client enables the investigator to upload and download files from Forensicloud. The investigator can upload tools that will be available to them and download analysis reports. The client will mirror the evidence not only to Forensicloud, but also to an image on the investigators computer. This eliminates the need of a second mirroring step, which speeds up the overall investigation time.

4.3 Investigation Component

Virtual machines provide investigators an environment for investigation of digital evidence. Investigators will use the same credentials used with the Forensicloud client to authenticate with this virtual machine. When an investigator creates a job, a fresh virtual machine is cloned from a base virtual machine. This virtual machine will have tools installed with licenses for the investigator to use. If the investigator needs a tool that is not already installed they can simply install it themselves. When an investigator is done analyzing evidence or when the job expires (jobs expire in 90 days by default), the virtual machine is deleted and all data corresponding to the job is deleted.

A virtualization manager is needed to control the creation and deletion of these virtual machines. Commercial software such as VMware ESXi and VMware vCenter [14] or open source software such as OpenStack [10] will fulfill all the needs of the investigation component. It will allow creation, cloning, modification, and deletion of virtual machines. It also handles virtual machine isolation [2]. In order for investigators to access virtual machines on-demand and remotely virtual desktop infrastructure (VDI) is needed. VDI provides on-demand access to a virtual machine that has either been provisioned upfront or as the investigator connects. It provides both input (keyboard, mouse, etc.) and output (monitor, sound, etc.). The particular VDI solution used is VMware Horizon View. VMware View provides the input and output as well as USB redirection, which allows a local USB device, such as a flash drive, to be connected to the remote virtual machine [1]. VMware Horizon View provides VDI by use of a standalone client that can authenticate with Forensicloud over a secure connection.

4.4 Processing Component

Forensic tools will be run using computing clusters. There are two options we are currently considering. The first is to use a virtualization cluster that has a similar architecture to that used in the investigation component; the second is to run the tools on an HPC cluster. Either solution has the ability to scale up or down based on the needs of the system and both will have measured service.

Using the virtualization cluster several virtual machines will handle the workload for the tools. The virtual machines will have an agent running that will accept payloads and process them with forensic tools. A standard blade server will contain one or more nodes (virtual machines). This option brings both advantages and disadvantages. Advantages are:

- virtual machines can run any operating system required by the particular tool being used
- they have dedicated resources
- the particular tool should not have to be modified to run on them
- a virtualization cluster can also be as big or as small as it needs to be since nodes of the cluster can be added and removed with little difficulty

There are also several disadvantages to using a virtualization cluster:

- virtualization overhead
- the need for a custom scheduler to schedule incoming jobs
- no default message passing interface (MPI)
- the need for a custom agent on each node to accept incoming jobs

Using an HPC cluster, a job is submitted to a scheduler node on the cluster and the other nodes of the cluster handle the work load. The cluster will have a scheduler for the jobs built in. Some of the benefits of using a cluster are:

- little overhead
- no custom scheduler
- MPI support

There are also a few disadvantages to using a HPC cluster:

- uses Linux for all the nodes; this means that all the tools running on the cluster must support Linux
- forensic tools may need to be modified to support the particular MPI used by the cluster
- nodes are more expensive than virtualization cluster nodes. However, each node of an HPC cluster will generally have superior hardware than the nodes of a virtualization cluster
- nodes are more specialized to handle computing jobs rather than general purpose jobs

Either solution can be used depending on the resources available. Figure 1 shows how each part of Forensicloud is connected to each other.

4.5 Investigation on Forensicloud

The process of investigation on Forensicloud follows three phases.

- Phase 1 – The investigator will specify the tools that will be ran on theevidence as well as other information about the case used for reporting

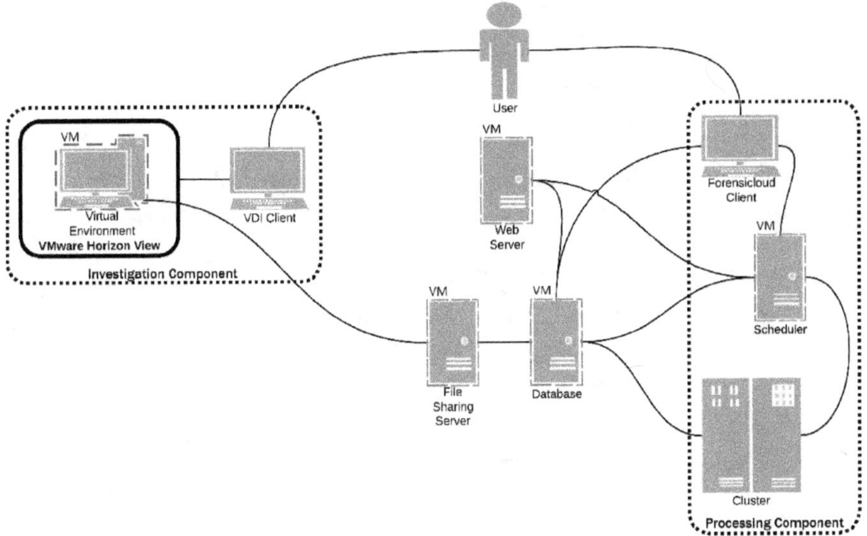

Figure 1 The components of Forensicloud

(case type, investigators, etc.). The investigator will upload any extratools they need to be available on the virtual machine used in phase 3. The investigator will then upload their evidence to Forensicloud using the Forensicloud client. Figures 2 and 3 illustrate the job setup and image uploading.

- Phase 2 – While the evidence is uploading specific tools that the investigator selected are run against the uploaded data. Only bitstream or single file capable tools will be run as the evidence is being uploaded. When the evidence has finished uploading the rest of the processing can be done on the evidence.
- Phase 3 – Using the investigation component, the investigator will be provided a virtual machine to examine the results returned from the processing component. They may also further analyze the evidence using

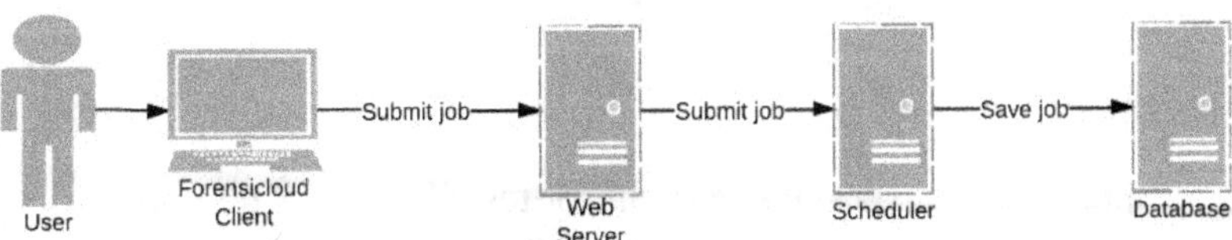

Figure 2 Phase 1 - Job setup

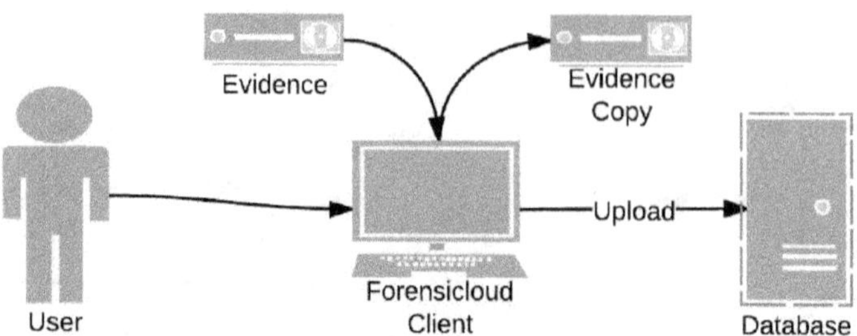

Figure 3 Phase 3 - Analysis retrieval

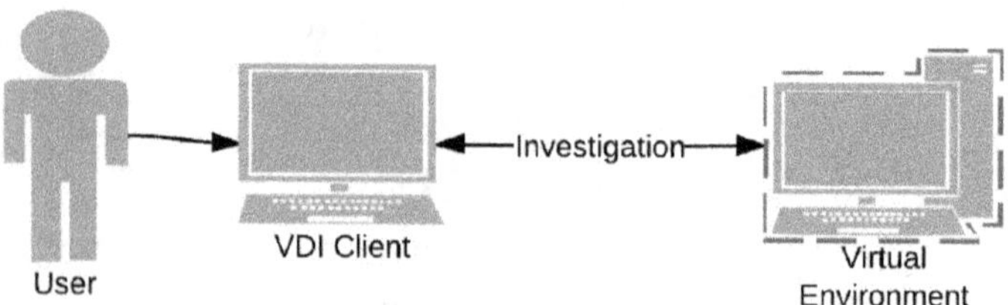

Figure 4 Phase 3 - Investigation

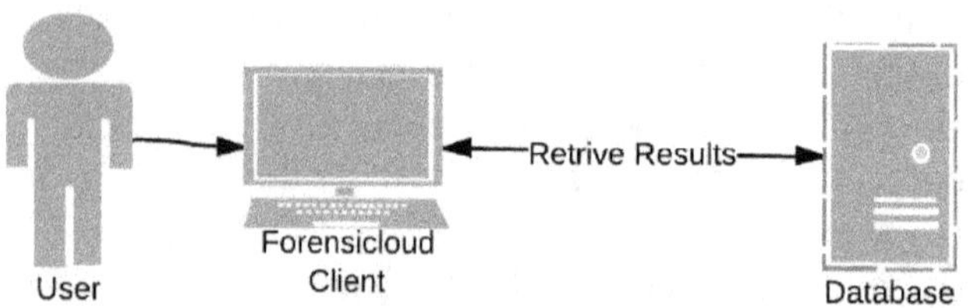

Figure 5 Phase 3 - Analysis retrieval

tools provided in the virtual machine as well as any tools they have uploaded. Figure 4 shows the investigation step. As the investigator performs analysis on the evidence they will be able to retrieve it for use in court. The retrieval of this data will be done using the Forensicloud client. Figure 5 displays the analysis retrieval part of phase 3.

5 Experiments

This section presents an experimental plan for testing various components of the Forensicloud architecture. For each phase of the process there are tests that should be conducted to determine the feasibility of the implementation. Some tests are also designed to establish benchmarks for the underlying components. Many of the files and images used in the following tests are from the digital corpora hosted by NPS [18].

Table 2 lists the systems that we are planning to use to test our implementation of Forensicloud. The HPC Cluster is named Talon and is a 34.4 TeraFLOPS IBM iDataPlex cluster with 256 nodes and is hosted by the High Performance Computing Collaboratory of Mississippi State University. The Virtualization Cluster nodes are running virtualization software so one node can act as several nodes. In the following tests each physical node has two virtual machines that are acting as nodes. This allows the virtualization cluster to have four nodes total with 12 threads and 24 GiB of RAM each. Table 2 lists the hardware specifications for one physical node.

5.1 Phase 1 Testing

Phase 1 encompasses setting up the initial processing tasks that should be performed on the evidence, the possible imaging of the evidence, and the encryption of the evidence for transmission across the Internet. All tests for this phase should be performed on a workstation as defined in Table 2 as all these tests relate to the Forensicloud client.

5.1.1 Simultaneous imaging and upload

Current research [31] [29] indicates that processing of data should start at the same time that the imaging process begins. Imaging and uploading, at least on Internet2, are I/O bound tasks when the target drive is a SATA HDD. Roussev et. al [31] report that a commodity SATA HDD can be read sequentially at 120 MB/s. This would take 2.5 hours to read a 1 TB drive. To facilitate faster processing we believe the client should support transmitting the data and creating the image in parallel.

Test – Determine the effect on performance of simultaneous imaging and network transmission of the evidence. The client should be run twice: once just to upload the test data to the server; once to upload the test data and

Table 2 Systems used in testing

Name	RAM	CPU(s)	Cores	Threads	Disk Speed
HPC Cluster	24 GiB	2 Intel Xeon X5660 @ 2.8 Ghz	12	24	100Mib/s
Virtualization Cluster	128 GiB	2 E5-2670 @ 2.6 GHz	16	32	1Gib/s
Workstation	8 GiB	Intel Core i7-860 @ 2.8 GHz	4	8	100Mib/s

make an image of it in parallel. This will be used to determine if the client can support both operations simultaneously while maintaining the optimal 120 MB/s upload rate.

5.1.2 Encryption

To ensure the confidentiality and integrity of the data during transmission, encryption must be applied to the evidence being transmitted between the client and server. The Advanced Encryption Standard (AES) is supported by NIST and is a well-known, reliable, symmetric key encryption protocol [13]. We consider this sufficient for the purpose of securing data during network transmission.

Test – Determine if AES encryption has any effect on the overall upload rate of the evidence being transmitted across the Internet. The client should be run twice: once just to upload test data to the server; once to encrypt the test data prior to transmission. If the upload rate is significantly affected by the encryption of the data, other reliable encryption standards may be explored.

5.2 Phase 2 testing

Phase 2 of Forensicloud is the processing of the digital evidence by the backend cluster. The goal of this phase is to efficiently process incoming data in order to provide the investigator with analyzable results as soon as possible. Testing in this phase relates to the number of nodes, or cores, necessary to carry various tasks within the constraints of the system (network transmission time, I/O speed, etc.).

5.2.1 Real-time stream processing

Streamlining the analysis process requires processing the data as it becomes available to the server. In [20] Garfinkel mentions two approaches to processing digital evidence: *bulk data analysis* and *file-based approaches. Bulk data analysis* processes blocks of data without regard to the underlying structure. That is, it does not require filesystem structural information in order to carry out its task. These types of techniques are ideal for processing the stream of data being transmitted from the Forensicloud client. Given that we expect the process to be I/O bound from the client and that SATA HDDs are still more common than solid state drives (SDDs), the transfer rate will likely be limited to 120 MB/s. This requires 120 – 200 cores to maintain processing at this speed [31]. We are currently aware of two tools designed for processing a stream of data: bulk extractor [34] and sdhash [30].

Test – Determine the required number of cores for a Forensicloud implementation to run bulk extractor and sdhash when data is being streamed from the client. Received data needs to be chunked and passed to each tool with nodes being allocated as needed.

File-based approaches process at the file level and many tools use this method. By extracting files from a stream in realtime, it is possible for the Forensicloud cluster to perform file level analysis. The latency-optimized target acquisition prototype mentioned in [31] shows it is possible to reassemble files from a stream in realtime and make them available to clients. In this case, the clients are the nodes. Once the files are available, they can be passed to appropriate tools for file-centric processing. While many tools operate on files, not all of them make sense to run during the initial processing. Password cracking tools for instance will only need to be run against a small subset of files on a given system and then with very specific configurations. Exif data however is something many file types possess that can be extracted in an automated manner.

Test – Determine required number of cores for ExifTool [22] to process files extracted from an uploaded data stream in realtime.

5.3 Phase 3 Testing

During phase 3 the investigator will connect to a remote virtual machine. They will use this virtual machine to perform analysis on the evidence and the reports generated by the various tools run on the cluster. As reports are made the investigator can use the Forensicloud client to retrieve them. The investigator can run certain tools on the cluster and others will be run on the virtual machine itself. Using the client they can also upload additional tools to use within the virtual machine.

5.3.1 Remote desktop connection

The investigator connects to a remote virtual machine by use of a remote desktop protocol that is built into VDI. If VMware Horizon View is used for VDI there are two protocol options: PCoIP and Microsoft's RDP. PCoIP has more features, such as USB redirection and multiple-monitor support.

Test – This test will determine the usability of the remote desktop protocol used. To determine this connect to Forensicloud using VDI from an average internet connection. Determine if the virtual machine is still usable from these locations by performing typical investigation activities, such as viewing reports and using forensic tools. The virtual machine should responsive and not have any noticeable video delay.

5.4 Performance Tests

To determine the optimal processing option for processing digital evidence in Forensicloud, it is important to first get a baseline of how fast a forensics *workstation* can process digital evidence. For each tool described below, the tool will use data from the digital corpora hosted by NPS [18] unless otherwise specified. For each tool, the amount of time that the tool takes to complete the processing will be used as a benchmark to determine the overall performance improvement of the distributed processing options.

After the benchmark for each tool has been set, each tool will be executed using the virtualization cluster and HPC cluster to determine the speeds that can be achieved when each tool is utilizing parallelization. Based on the speeds of each tool in the different environments, it can be determined which distributed processing solution is most useful.

Test – Bulk Extractor – the goal of this test is to compare HPC cluster versus virtualization cluster performance of Bulk Extractor [34]. Using a disk image of 1 TiB or greater, break the image into overlapping fragments. The number of fragments will be equal to the number of nodes and they will be of equal size. Write the output data back to the file server. Record and analyze processing time for each environment.

Test – sdhash – the goal of this test is to compare HPC cluster versus virtualization cluster performance of sdhash [30]. Using a disk image of 1 TiB or greater, break the image into overlapping fragments. The number of fragments will be equal to the number of nodes and they will be of equal size. Write the output data back to the file server. Record and analyze processing time for each environment.

Test – Password Cracking with John the Ripper – the goal of this test it to compare HPC cluster versus virtualization cluster performance of John the Ripper [28]. Construct and store Windows NTLM hashes corresponding to password lengths of 4, 8, 12, and 16. Run John the Ripper will using MPI on the HPC cluster, using its 'node' option on the virtualization server, and using only threading on the forensics workstation. Record and analyze processing time for each environment.

Test – Extract metadata from files with ExifTool – the goal of this test is to compare HPC cluster versus virtualization cluster performance of ExifTool [22]. Store the Govdocs1 corpus files on the file server. For each test run multiple ExifTool instances at the same time. The number of instances equals the number of threads per node multiplied by the number of nodes used. Record and analyze processing time for each environment.

Test – Extract artifacts from Windows memory dump using Volatility – the goal of this test is to compare HPC cluster versus virtualization cluster performance of Volatility [15]. Create memory images of size 4 GiB, 8 GiB, and 16 GiB. Run a single Volatility command on each node. Record and analyze processing time for each environment.

5.5 Other Testing

This section contains tests that should be performed that do not directly test Forensicloud. However, these tests should be performed to ensure Forensicloud will not be limited by external complications.

5.5.1 Nodenize

Nodenize is a tool made for Forensicloud that allows forensic tools that only work with single input files work in parallel. Nodenize takes as input a directory containing files to be processed. It also takes a tool that will be used to process files in the directory as input. Nodenize will be run on each node used for the task; each of these nodes will know what set of the input files it will process by using a node identifier. If the identifier is 1 out of 4 the node will process the first 25% of the data, if it is 2 out of 4 will process the second 25% of the data, etc. It then gets a list of all the files in the directory and determines which section of them it will process. Finally, it processes each file using the tool selected.

Test – The goal of this test is to determine if nodenize has any negative influence on the performance of forensic tools ran with it. To test this run a tool with nodenize and run the same tool without nodenize. Running the tool without nodenize will require batch execution of the tool on the different input. Determine if nodenize had any noticeable influence on performance.

5.5.2 Workload

There are many areas of in a Forensicloud workflow that will need to be tested to establish the most efficient and effective use of that specific Forensicloud implementation. These tests layout procedures that should be used to determine the capability and processing load that a specific Forensicloud implementation can handle.

Test – Single Upload and Download Capability – the goal of this test is to determine upload capability to a Forensicloud environment. Using varying data set sizes, upload and download each data set individually from different Forensicloud facilities. Record and document how much time it took to upload each file. Determine the amount of time that would prohibitive for a single user.

Test – Multiple Upload and Download Capability – the goal of this test is to determine multiple upload capability to a Forensicloud environment. Using a set data set size of 500GB or greater, upload and download a data set from 2 Forensicloud facilities at the same time. Record how much time it took to upload each file. Increment the upload locations by 1 and upload the files again. Record how much time it took to upload each file. Continue adding one facility until the time for upload would be prohibitive to any user.

Test – Nodes per Task – the goal of this test is to determine the number of nodes that is optimal for each task. Using a data set of 1TiB or greater run each tools available to the user. With every run of the tool add 1 for the tool to use. Continue this until there is either no more nodes remaining or there is no longer an increases level of performance by adding another node. Document the optimal number of nodes for each task.

Test – Nodes per case – the goal of this test is to determine the number of nodes for each user of Forensicloud. Starting with a single case, allocate all nodes to all cases equally. Increment the number of cases by one until the processing performance is no longer optimal. Record the number of nodes a single case needs to be effective.

6 Preliminary Test Results

We were unable to perform a full evaluation of the system due to various technical difficulties. However, one test was completed that tests various aspects of the particular cluster being used. Bulk Extractor was run on a forensic workstation, virtualization cluster, and HPC cluster. Table 3 displays the evidence used in the tests. The first test used only one node of each cluster; the second test used bulk extractor's parallelization option '-Y' to separate the evidence into four chunks. Each chunk was processed by one node of each cluster. Bulk Extractor was run with all default scanners active. Table 4 displays the results of this test with only one node per cluster. The time and processing rate reported for node of each cluster is shown Tables 5–7. These tables also display whether Bulk Extractor thought the process was CPU or I/O bound.

Table 3 Details of the evidence used in testing

	ubnist1.gen3.E01	nps-2009-domexusers.E01
Compressed	854 MiB	4.07 GiB
Uncompressed	1.96 GiB	40 GiB

Table 4 Results of one node per cluster

	ubnist1.gen3.E01			nps-2009-domexusers.E01		
	HPC	Virtualization	Workstation	HPC	Virtualization	Workstation
Seconds	112.3	44.8	101.0	892.1	566.4	894.2
MiB/s	18.75	47.0	20.9	48.2	75.8	48.0
MiB	2106	2106	2106	42949	42949	42949
Bound	CPU	CPU	CPU	CPU	None	CPU
Threads	12	12	8	12	12	8
RAM	24	24	8	24	24	8

Table 5 The workstation cluster results. Each workstation had 8 GiB RAM and 8 threads

	ubnist1.gen3.E01				nps-2009-domexusers.E01			
	Node 1	Node 2	Node 3	Node 4	Node 1	Node 2	Node 3	Node 4
Seconds	35.9	44.9	26.0	5.1	462.3	203.5	196.0	47.9
MiB/s	14.5	11.6	20.0	107.4	23.2	52.7	54.8	224.1
MiB	520	520	520	546	10737	10737	10737	10737
Bound	CPU	CPU	CPU	None	CPU	CPU	CPU	I/O

Table 6 The HPC cluster results. Each node had 24 GiB RAM and 12 threads

	ubnist1.gen3.E01				nps-2009-domexusers.E01			
	Node 1	Node 2	Node 3	Node 4	Node 1	Node 2	Node 3	Node 4
Seconds	45.6	64.1	31.3	10.9	480.6	260.2	165.1	53.4
MiB/s	11.4	8.1	16.6	50.1	22.3	41.3	65.0	201.0
MiB	520	520	520	546	10737	10737	10737	10737
Bound	CPU	CPU	CPU	None	CPU	CPU	CPU	I/O

Table 7 The virtualization cluster results. Each node had 24 GiB RAM and 12 threads

	ubnist1.gen3.E01				nps-2009-domexusers.E01			
	Node 1	Node 2	Node 3	Node 4	Node 1	Node 2	Node 3	Node 4
Seconds	20.2	26.1	14.3	6.1	240.3	171.3	163.7	47.5
MiB/s	25.8	19.9	36.3	90.3	44.7	62.7	65.6	225.9
MiB	520	520	520	546	10737	10737	10737	10737
Bound	None	None	None	None	CPU	None	CPU	I/O

The HPC cluster performed similarly to the workstation. We believe that the primary issue with our HPC cluster is that is currently lacks a fast storage device. Due to technical difficulties we are unable to provide it with the same or similar storage device that is used by the virtualization cluster. If faster storage was used it would improve the processing speed. The virtualization cluster outperformed both the workstation and the HPC cluster by up to 225%. Further testing needs to be conducted with a shared storage device. Even though the HPC cluster and workstation cluster performed similarly, we believe the HPC cluster is still more economical and feasible than a workstation cluster. It would take 384 workstations to reach the 3072 threads in the HPC cluser we used. Thread for thread, it would take 96 virtualization nodes to equal our HPC cluster's threads.

7 Future Work

There is still significant work to be done to fully implement a working prototype of Forensicloud. One goal of this paper is to enumerate the challenges Forensicloud faces in order to start a conversation within the community. This will provide us useful feedback that we can incorporate into the prototype we are developing.

Only a small subset of tools were discussed in this paper, we will complete a comprehensive review of digital forensics tools to determine which tools can be parallelized appropriately for use in Forensicloud. This includes identifying those tools that operate on discrete elements such as files and those that are designed to be executed in parallel.

An aspect of Forensicloud we have not mentioned in this paper is its use as Platform as a Service (PaaS). If tool developers have access to a Forensicloud they can use it to build and test parallelized tools for digital forensics. We will implement an API that will allow investigators to easily interface with Forensicloud. With this API a tool can be made that utilizes parallel processing without the need to write parallel code.

8 Conclusion

The quantity and diversity of digital evidence continues to increase. What was once a matter of investigating thousands of files has turned into investigating millions of files of various types. An investigator can no longer manually search evidence; smarter tools are required that can automate as much of the process as possible. These tools work on traditional forensics workstations; however, they can take hours or even days to finish on larger evidence. Forensicloud decreases the time needed to process data by leveraging the power of a high performance computing platform and by adapting existing tools to operate within this environment. Forensicloud further improves the investigation by providing an environment for remote investigations that gives investigators access to licensed tools, such as Access Data's FTK and Guidance Software's Encase that they may not have had before.

In this paper we have presented an architecture for a cloud-based digital forensics analysis platform. A Forensicloud system provides several benefits:

- it reduces the overall processing time of large quantities of data by leveraging the power of a high performance computing platform and adapting existing tools to operate within this environment
- it provides smaller departments that may not have access to certain commercial software the ability to use this software remotely
- it enables collaboration. With the evidence stored in the cloud, it is possible to allow someone to access the case, provided sufficient authorization from the case owner, to provide additional analysis. An attorney could also have access to the case on the cloud to view the investigators analysis.

We have detailed several challenges we believe an implementation of Forensicloud must overcome in order to gain acceptance from both the judicial and technical communities. We have presented guidelines that address these challenges based on existing standards, where applicable.

Finally, we have presented a test plan for evaluating various components of a Forensicloud implementation. Using these tests one can determine the feasibility of the architecture for the particular implementation of Forensicloud.

References

[1] Vdi: A new desktop strategy. Technical report, VMware Inc., Palo Alto, CA, 2006.

[2] vsphere security esxi 5.1. Technical report, VMware Inc., Palo Alto, CA, 2012.

[3] Citrix xenserver. http:www.citrix.comproductsxenserveroverview.html, 2014. Accessed: 2014-07-20.

[4] Encase forensic. https:www.guidancesoftware.comproductsPagesencase-forensicoverview.aspx, 2014. Accessed: 2014-07-20.

[5] Forensic tookkit. http:www.accessdata.comsolutionsdigitalforensicsftk, 2014. Accessed: 2014-07-20.

[6] Household upload index - united states. http:www.netindex.comupload2, 1United-States, 2014. Accessed: 2014-07-21.

[7] Kvm. http:www.linux-kvm.orgpageMain Page, 2014. Accessed: 2014-07-20.

[8] Microsoft hyper-v. http:www.microsoft.comen-usserver-cloudsolutions virtualization.aspx, 2014. Accessed: 2014-07-20.

[9] Mississippi optical network. http:mission.mississippi.edu, 2014. Accessed: 2014-07-25.

[10] Openstack. http:www.openstack.org, 2014. Accessed: 2014-07-20.

[11] Openvz. http:openvz.orgMain Page, 2014. Accessed: 2014-07-20.

[12] Sleuth kit hadoop. http://www.sleuthkit.org/tsk hadoop/, 2014. Accessed: 2014-07-20.

[13] Standards and guidelines tested under the cavp.http:csrc.nist.govgroups STMcavpstandards.html, 2014. Accessed: 2014-07-27.

[14] Vmware esxi. http:www.vmware.comproductsvspherehypervisor, 2014. Accessed: 2014-07-20.

[15] The volatility framework 2.31. https:code.google.compvolatility, 2014. Accessed: 2014-07-27.

[16] Xen project. http:www.xenproject.org, 2014. Accessed: 2014-07-20.

[17] Welcome to apache hadoop. http:www.hadoop.apache.org, (Accessed July 20 2014).

[18] Simson Garfinkel, Paul Farrell, Vassil Roussev, and George Dinolt. Bringing science to digital forensics with standardized forensic corpora. *digital investigation*, 6:S2–S11, 2009.

[19] Simson L Garfinkel. Digital forensics research: The next 10 years. *Digital Investigation*, 7:S64–S73, 2010.

[20] Simson L Garfinkel. Digital media triage with bulk data analysis and bulk_extractor. *Computers & Security*, 32:56–72, 2013.

[21] George Grispos, Tim Storer, and W Glisson. Calm before the storm: The challenges of cloud computing in digital forensics. *International Journal of Digital Crime and Forensics*, 4(2):28–48, 2012.

[22] Phil Harvey. Exiftool 9.69. http:www.sno.phy.queensu.ca$\tilde{p}$hilexiftool, 2014. Accessed: 2014-07-27.

[23] Hanan Hibshi, Timothy Vidas, and Lorrie Faith Cranor. Usability of forensics tools: a user study. In *IT Security Incident Management and IT Forensics (IMF), 2011 Sixth International Conference on*, pages 81–91. IEEE, 2011.

[24] Peter Mell and Tim Grance. The nist definition of cloud computing. 2011.

[25] Scientific Working Group on Digital Evidence. Swgde model quality assurance manual for digital evidence laboratories, 2012.

[26] Marc Parisi, David A Dampier, Rayford Vaughn, and Yoginder Dandass. Improving foremost execution speed by data and task level parallelization. 2009.

[27] Nicole Perlroth. Tally of cyber extortion attacks on tech companies grows. http:bits.blogs.nytimes.com20140619tallyofcyberextortionattack sontechcompanies-grows?php=true& type=blogs&r=0,Accessed: 2014-07-20.

[28] Openwall Project. John the ripper 1.7.9 jumbo 7. http:www.openwall.com john, 2014. Accessed: 2014-07-27.

[29] Vassil Roussev. Scalable data correlation. In *Eighth annual IFIP WG*, volume 11, 2012.

[30] Vassil Roussev. sdhash 3.4. http:roussev.netsdhashsdhash.html, 2014. Accessed: 2014-07-27.

[31] Vassil Roussev, Candice Quates, and Robert Martell. Real-time digital forensics and triage. *Digital Investigation*, 10(2):158–167, 2013.

[32] Vassil Roussev and Golden G Richard III. Breaking the performance wall: The case for distributed digital forensics. In *Proceedings of the 2004 Digital Forensics Research Workshop*, volume 94, 2004.

[33] Vassil Roussev, Liqiang Wang, Golden Richard, and Lodovico Marziale. A cloud computing platform for large-scale forensic computing. In *Advances in Digital Forensics V*, pages 201–214. Springer, 2009.

[34] Naval Postgraduate School. bulk extractor 1.5 alpha 6. https:github.comsi msongbulk_extractor, 2014. Accessed: 2014-07-27.

Biographies

Cody Miller is a Research Associate for the Distributed Analytics and Security Institute at Mississippi State University. Cody's research interests are in Cloud Computing, Computer Security, and Digital Forensics. He has a B.S. and M.S. Degree in Computer Science & Engineering from Mississippi State University. In his graduate studies he worked for the National Forensics Training Center at Mississippi State University where he taught law enforcement officers digital forensics.

Dae Glendowne is an Assistant Research Professor at the Distributed Analytics Security Institute at Mississippi State University. He is currently pursuing his Ph.D. in Computer Science at Mississippi State University. He has a B.S. Degree in Computer Science from the University of Tennessee at Martin and an M.S. Degree in Computer Science from Mississippi State University. His research interests include malware analysis, memory forensics, and applying machine learning to computer security problems.

Dr. Dave Dampier is a Professor of Computer Science & Engineering at Mississippi State University specializing in Digital Forensics and Information Security. He currently serves as Director of the Distributed Analytics and Security Institute, the university level research center charged with Cyber Security Research. In his current capacity, Dr. Dampier is the university lead for education and research in cyber security. Prior to joining MSU, Dr. Dampier spent 20 years active duty as an Army Automation Officer. He has a B.S. Degree in Mathematics from the University of Texas at El Paso, and M.S. and Ph.D. degrees in Computer Science from the Naval Postgraduate School. His research interests are in Cyber Security, Digital Forensics and Software Engineering.

Kendall Blaylock received his M.S. and B.S. degrees from Mississippi State University. During that time he worked as a research assistant in the area of computer forensics. After graduating from MSU he then went on to work for the National Forensic Training Center at MSU. At the NFTC Kendall is currently serving as a Research Associate III. The research associate position at the NFTC requires Kendall to be an instructor as well as a researcher in the area of digital forensics. As an instructor for the NFTC, Kendall provides training for law enforcement officers and Military Veterans. In addition to being an instructor for the NFTC, he also oversees and conducts research projects at the NFTC. These projects are intended to benefit the digital forensics community and allow law enforcement to conduct investigations in a more effective and efficient manner. Kendall's background in the College of Business at MSU enables him to research where digital forensics is involved with business operations, such as the area of e-discovery and internal corporate investigation.

Structure Preserving Large Imagery Reconstruction

Ju Shen[1], Jianjun Yang[2], Sami Taha-abusneineh[3], Bryson Payne[4] and Markus Hitz[4]

[1]*Department of Computer Science, University of Dayton, 300 College Park, Dayton, OH 45469, USA*
[2]*Department of Computer Science and Information Systems, University of North Georgia, Oakwood, GA 30566, USA*
[3]*Computer Science Department, Palestine Polytechnic University (PPU), Ein Sara Street, Hebron, Palestine*
[4]*Department of Computer Science and Information Systems, University of North Georgia, Dahlonega, GA 30597, USA*
jshen1@udayton.edu, jianjun.yang@ung.edu, staha77@yahoo.com, {bryson.payne; markus.hitz}@ung.edu

Received 28 May 2014; Accepted 20 August 2014
Publication 7 October 2014

Abstract

With the explosive growth of web-based cameras and mobile devices, billions of photographs are uploaded to the internet. We can trivially collect a huge number of photo streams for various goals, such as image clustering, 3D scene reconstruction, and other big data applications. However, such tasks are not easy due to the fact the retrieved photos can have large variations in their view perspectives, resolutions, lighting, noises, and distortions. Furthermore, with the occlusion of unexpected objects like people, vehicles, it is even more challenging to find feature correspondences and reconstruct realistic scenes. In this paper, we propose a structure-based image completion algorithm for object removal that produces visually plausible content with consistent structure and scene texture. We use an edge matching technique to infer the potential structure of the unknown region. Driven by the estimated

Journal of Cyber Security, Vol. 3, 263–288.
doi: 10.13052/jcsm2245-1439.332

structure, texture synthesis is performed automatically along the estimated curves. We evaluate the proposed method on different types of images: from highly structured indoor environment to natural scenes. Our experimental results demonstrate satisfactory performance that can be potentially used for subsequent big data processing, such as image localization, object retrieval, and scene reconstruction. Our experiments show that this approach achieves favorable results that outperform existing state-of-the-art techniques.

1 Introduction

In the past few years, the massive collections of imagery on the Internet have inspired a wave of work on many interesting big data topics. For example, by entering a keyword, one can easily download a huge number of photo streams related to it. Moreover, with the recent advance in image processing techniques, such as feature descriptors [1], pixel-domain matrix factorization approaches [2–4] or probabilistic optimization [5], images can be read in an automatic manner rather than relying on the associated text. This leads to a revolutionary impact to a broad range of applications, from image clustering or recognition [6–12] to video synthesis or reconstruction [13–15] to cyber-security via online images analysis [16–19] to other scientific applications [20–24].

However, despite the numerous applications, poor accuracy can be yielded due to the large variation of the photo streams, such as resolution, illumination, or photo distortion. In particular, difficulties arise when unexpected objects present on the images. Taking the Google street view as an example, the passing vehicles or walking passengers could affect the accuracy of image matching. Furthermore such unwanted objects also introduce noticeable artifacts and privacy issue in the reconstructed views.

To resolve these issues, object removal [25, 26] is an effective technique that has been widely used in many fields. A common approach is to use texture synthesis to infer missing pixels in the unknown region, such as [27, 28]. Efros and Leung [29] use a one-pass greedy algorithm to infer the unknown pixels based on an assumption that the probability distribution of a target pixel's brightness is independent from the rest of the image given its spatial neighborhood. Some studies propose example-based approaches to fill the unknown regions, such as [27–29]. These approaches failed to preserve the potential structures in the unknown region. Bertalmio *et al.* [30] apply *partial differential equations* (PDE) to propagate image Laplacians. While the potential structures are improved in the filled region, it suffers from blurred synthesized texture. Drori *et al.* [31], propose an enhanced algorithm to improve the rendered texture. Jia *et al.* [32] propose an texture-segmentation based approach using tensor-voting to achieve the same goal. But their approaches are computationally expensive. A widely used image in-painting technique developed by Criminisi *et al.* [26] aims to fill the missing region by a sequence of ordered patches by using the proposed confidence map. The priority of each patch is determined by the edge strength from the surrounding region. However, the potential structures in the in-painted region can not be well preserved, especially for those images with salient structures. The authors Sun *et al.* in [33] make an improvement through structure propagation, while this approach requires additional user intervention and the results may depend on the individual animators.

As an extension of our early work [34], we propose an automatic object removal algorithm for scene completion, which would benefit large imagery processing. The cue of our method is based on the structure and texture consistency. First, it predicts the underlying structure of the occluded region by edge detection and contour analysis. Then structure propagation is applied to the region followed by a patch-based texture synthesis. Our proposed approach has two major contributions. First, given an image and its target region, we develop an automatic curve estimation approach to infer the potential structure. Second, an orientated patch matching algorithm is designed for texture propagation. Our experiments demonstrate satisfactory results that outperform other techniques in the literature.

The rest of the paper is organized as follows: in Section 2, we give an example to demonstrate the basic steps of our image completion algorithm. Then we define the model and notations in Section 3. Details are further explained in Section 4. The experiment results are presented in Section 5. Finally we conclude the paper and our future work in Section 6.

2 A Simple Example

The process of our framework is: for a given image, users specify the object for removal by drawing a closed contour around it. The enclosure is considered as the unknown or target region that needs to be filled by the remaining region of the image. Figure 1(a) shows an example: the red car is selected as the removing object. In the resulting image, Figure 1(b), the occluded region is automatically recovered based on the surrounding available pixels.

Our algorithm is based on two observations: spacial texture coherence and structure consistency along the boundaries between the target and source

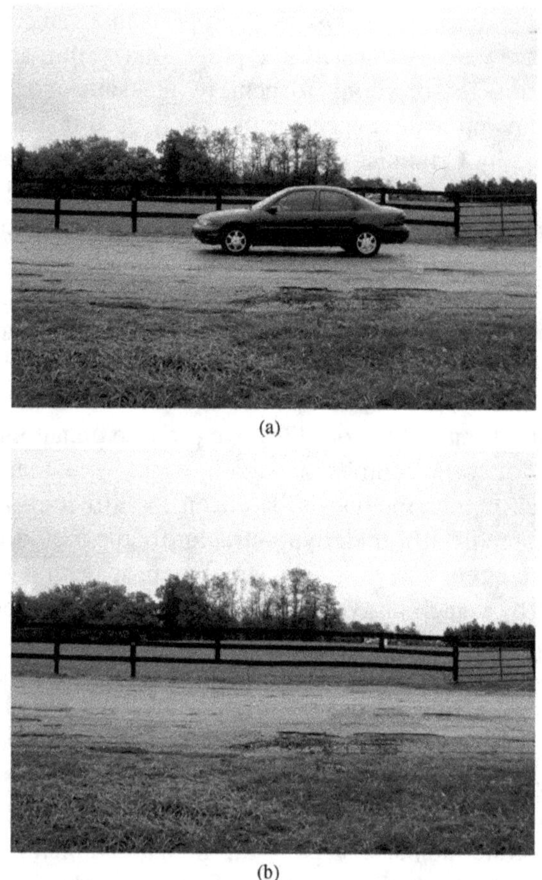

(a)

(b)

Figure 1 Example object removal on an image (a) The original image (b) The result image

regions. To ensure spacial coherence, many exemplar-based methods have been proposed to find the potential source texture for the target region. By traversing the available pixels from the known region, a set of "best patches" are found to fill the target region. Here the definition of "best patch" refers to a small region of contiguous pixels from the source region that can maximize a certain spacial coherence constraint specified by different algorithms. A typical example can be found in [26]. However, a naive copy-and-paste of image patches may introduce noticeable artifacts, though the candidate patches can maximize a local coherence. To resolve this problem, structure preservation is considered to ensure the global consistency. There have been several techniques presented for structure propagation to ensure smooth and natural transitions among salient edges, such as the Sun's method [33], which requires additional user input to finish the task.

3 The Approach

First let us define some notations for the rest of paper. The target region for filling is denoted as Ω; the remaining part $\Phi(= \mathrm{I} - \Omega)$ is the region whose pixels are known. The boundary contours are denoted as $\partial\Omega$ that separate Φ and Ω. A pixel's value is represented by $p = \mathbf{I}(\mathrm{x, y})$, where x and y are the coordinates on the image **I**. The surrounding neighborhood centered at *(x, y)* is considered as a patch, denoted by Ψ_p, whose coordinates are within $[x \pm \Delta x, y \pm \Delta \mathrm{y}]$, as shown in Figure 2. In our framework, there are three stages involved: structure estimation, structure propagation, and remaining part filling.

Structure Estimation: In this stage, we estimate the potential structures in the target region Ω. To achieve this, we apply *gPb Contour Detector* [35] to extract the edge distribution on the image:

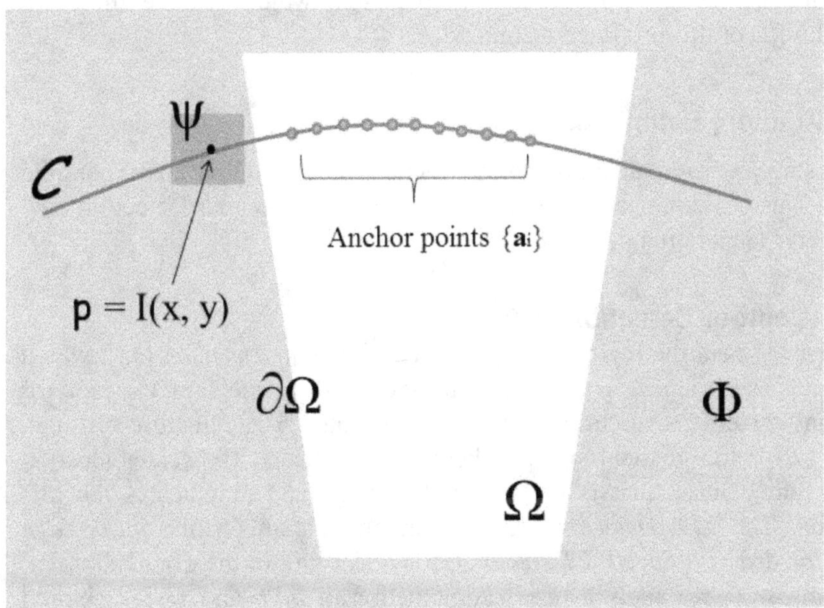

Figure 2 Symbols definition

$$I_{edge} = \sqrt{[\frac{\partial}{\partial x}(I * G_x)]^2 + [\frac{\partial}{\partial y}(I * G_y)]^2} \quad (1)$$

where, G_x and G_y are the first derivative of Gaussian function with respect to x and y axis ($G_x = \frac{-x.G(x,y)}{\sigma^2}$). After computing I_{edges}, most of the strong edges can be extracted via threshold suppression. Inspired by the level lines technique [36], the edges in Ω can be estimated by linking matching pairs of edges along the contour.

Structure Propagation: After the structures are estimated, textures along the structures are synthesized and propagated into the target region $\boldsymbol{\Omega}$. We use Belief Propagation to identify optimal patches of texture from the source region Φ and copy and paste them to the structures in Ω.

Remaining Part Filling: After the structure propagation, the remaining unfilled regions in Ω are completed. We adopt the Criminisi's method [26], where a priority-based patch filling scheme is used to render the remaining target region in an optimal order.

In the following subsections, we present the details of each step of the proposed algorithm. In particular, we give emphasis to the first two steps: structure estimation and structure propagation, which provide the most contribution of this proposed technique.

3.1 Structure Estimation

In this stage, we estimate the potential structures in Ω by finding all the possible edges. This procedure can be divided into two steps: Contour Detection in Φ and Curve Generation in Ω.

3.1.1 Contour Detection in Φ

We first segment the region Φ by using *gPb Contour Detector* [35], which computes the oriented gradient signal $G(x, y, \theta)$ on the four channels of its transformed image: brightness, color *a*, color *b* and texture channel. $G(x, y, \theta)$ is the gradient signal, where (x, y) indicates the center location of the circle mask that is drawn on the image and θ indicates the orientation. The *gPb Detector* has two important components: *mPb Edge Detector* and *sPb Spectral Detector* [35]. According to the gradient ascent on F-measure, we apply a linear combination of *mPb* and *sPb* (factored by β and γ):

$$gPb(x, y, \theta) = \beta \cdot mPb(x, y, \theta) + \gamma \cdot sPb(x, y, \theta) \qquad (2)$$

Thus a set of edges in Φ can be retrieved via *gPb*. However, these edges are not in close form and have classification ambiguities. To solve this problem, we use the *Oriented Watershed Transform* [35] and *Oriented Watershed Transform* [37] (*OWT-UCM*) algorithm to find the potential contours by segmenting the image into different regions. The output of *OWT-UCM* is a set of different contours $\{C_i\}$ and their corresponding boundary strength levels $\{L_i\}$.

3.1.2 Curve Generation in Ω

After obtaining the contours $\{C_i\}$ from the above procedure, salient boundaries in Φ can be found by traversing $\{C_i\}$. Our method for generating the curves in Ω is based on the assumption: for the edges on the boundary in Φ that intersects with the $\partial\Omega$, it either ends inside Ω or passes through the missing region Ω and exits at another point of $\partial\Omega$. Below is our algorithm for identifying the curve segments in Ω:

Algorithm 3.1 Identifying curve segments in Ω

Require: Reconstruct missing curves segments in Ω
Ensure: The estimated curves provide smooth transitions between edges in Φ

Initial t = 1.0
For t = t −Δt
    if $\exists$e $\in \{C\} : E \cap \partial\Omega \neq \emptyset$
        Insert *e* into $\{E\}$
End if t < δ_t
Set t = t_0, retrieve all the contours in $\{C_i\}$ with $L_i >$ t
Obtain $< \phi_{x1}, \phi_{x2} >$ for each E_x
DP on $\{< \phi_{01}, \phi_{02} >, < \phi_{11}, \phi_{12} >, \ldots\}$ to find optimal pairs
According to the optimal pairs, retrieve all the corresponding edge-pairs: $\{(E_{x1}, E_{x2}), (E_{x3}, E_{x4}), \ldots)\}$.
Compute a transition curve C_{st} for each $\{(E_s, E_t)\}$.

In algorithm 3.1, it has three main parts: (a) collect all potential edges $\{E_x\}$ in Φ that hits $\partial\Omega$; (b) identify optimal edge pairs $\{(E_s, E_t)\}$ from $\{E_x\}$; (c) construct a curve C_{st} for each edge pair (E_s, E_t).

Edges Collection: The output of *OWT-UCM* are contours sets $\{C_i\}$ and their corresponding boundary strength levels $\{L_i\}$. Given different thresholds t, one can remove those contours C with weak L. Motivated by this, we use the *Region-Split* scheme to gradually demerge the whole Φ into multiple subregions and extract those salient curves. This process is carried out on lines 1–9: at the beginning the whole region Φ is considered as one contour; then iteratively decrease t to let potential sub-contours $\{C_i\}$ faint out according the boundary strength; Every time when any edges *e* from the newly emerged contours $\{C\}$ were detected of intersecting with $\partial\Omega$, they are put into the set $\{E\}$.

Optimal Edge Pairs: the motivation of identifying edge pairs is based on the assumption if an edge is broken up by Ω, there exists a pair of corresponding contour edges in Φ that intersect with $\partial\Omega$. To find the potential pairs $\{(E_s,\ E_t)\}$ from the edge list $\{E_x\}$, we measure the corresponding enclosed regions similarities. The neighboring subregions $<\phi_{x1}^{(s)}, \phi_{x2}^{(s)}>$ which are partitioned by the edge E_s are used to compare with the corresponding subregions $<\phi_{x3}^{(s)}, \phi_{x4}^{(s)}>$ of another edge E_t. This procedure is described on lines 7 – 9 of the algorithm 3.1. For simplicity, the superscripts (*s*) and (*t*) are removed and the neighboring subregions $<\phi_{x1}, \phi_{x2}>$ are list in a sequential order. Each neighboring region is obtained by lowing down the threshold value t to faint out its contours as Figure 3 shows.

Figure 3 Contour extraction by adjusting the value of **t**

To compute the similarity between regions, we use the *Jensen-Shannon divergence* [38] method that works on the color histograms:

$$d(H_1, H_2) = \sum_{i=1}^{n} \{H_1^i \, . \, log \frac{2.H_1^i}{H_1^i + H_2^i} + H_2^i.log \frac{2.H_1^i}{H_1^i + H_2^i}\} \tag{3}$$

where H_1 and H_2 are the histograms of the two regions ϕ_1, ϕ_2; i indicates the index of histogram bin. For any two edge *(E_s, E_t)*, the similarity between them can be expressed as:

$$M(E_s, E_t) = \frac{\|L_s - L_t\|}{L_{max}} . \min\{d(H_{si}, H_{ti}) + d(H_{sj}, H_{tj})\} \tag{4}$$

i and *j* are the exclusive numbers in $\{1, 2\}$, where 1 and 2 represent the indices of the two neighboring regions in Φ around a particular edge. The L_{max} is the max value of the two comparing edges' strength levels. The first multiplier is a penalty term for big difference between the strength levels of the two edges. To find the optimal pairs among the edge list, dynamic programming is used to minimize the global distance: $\sum_{s,t} M(E_s, E_t)$, where $s \neq t$ and *s, t* $\in \{0, 1, ..., size(\{E_i\})\}$. To enhance the accuracy, a maximum constraint is used to limit the regions' difference: $d(H_1, H_2) < \delta_H$. If the individual distance is bigger than the pre-specified threshold δ_H, the corresponding region matching is not considered. In this way, it ensures if there are no similar edges existed, no matching pairs would be identified.

Generate Curves for each *(E_s, E_t)***:** we adopt the idea of fitting the clothoid segments with polyline stoke data first before generating a curve [39]. Initially, a series of discrete points along the two edges E_s and E_t are selected, denoted as $\{p_{s0}, p_{s1}, ..., p_{sn}, p_{t1}, ..., p_{tm}\}$. These points have a distance with each other by a pre-specified value Δ_d. For any three adjacent points $\{p_{i-1}, p_i, p_{i+1}\}$, the corresponding curvature k_i could be computed according to [40]:

$$K_i = \frac{2 \, . \, det(p_i - p_{i-1}, p_{i+1} - p_i)}{||p_i - p_{i-1}||.||p_{i+1} - p_i||.||p_{i+1} - p_{i-1}||} \tag{5}$$

Combining the above curvature factors, a sequence of polyline are used to fit these points. The polylines are expected to have a possibly small number of line segments while preserving the minimal distance against the original data. Dynamic programming is used to find the most satisfied polyline sequence by giving a penalty for each additional line segment. A set of clothoid segments

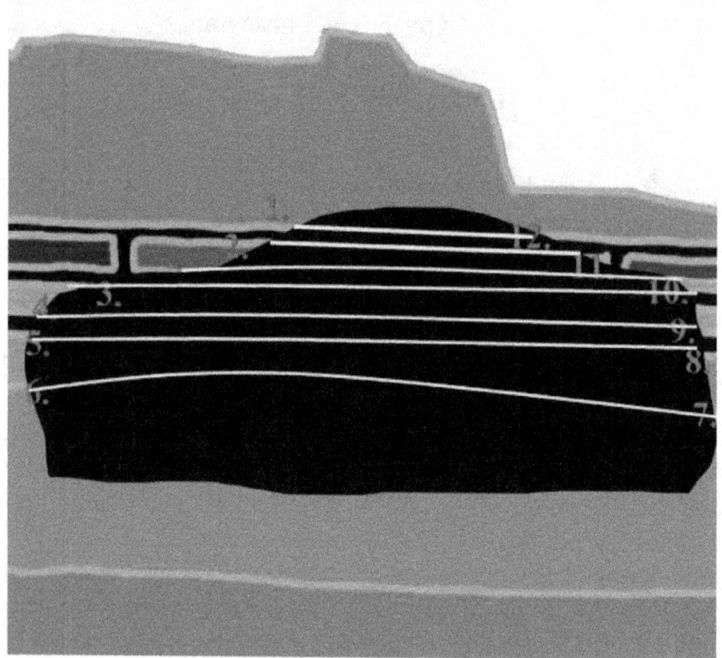

Figure 4 Optimal structure generation in Ω

can be derived corresponding to each line segment. After a series rotations and translations over the clothoid, a final curve C is obtained by connecting each adjacent pair with G^2 continuity [39]. Figure 4 demonstrates the curve generation result.

3.2 Structure Propagation

After the potential curves are generated in Ω, a set of texture patches, denoted as $\{\Psi_0, \Psi_1, ...\}$, need to be found from the remaining region Φ and placed along the estimated curves by overlapping with each other with a certain proportion. Similar to the Sun's method [33], an energy minimization based Belief Propagation(BP) framework is developed. We give different definitions for the energy minimization and passing messages, the details of which can be found in algorithm 3.2.

In our algorithm, the *anchor points* are evenly distributed along the curves with an equal distance from each other Δd. These points represent the center where the patches $\{\Psi_i\}(l \times l)$ are placed, as shown in Figure 2. In practice, we define $\Delta d = \frac{1}{4} \cdot l$. The $\{\hat{\Psi}_t\}$ is the source texture patches in Φ. They are chosen on from the neighborhood around $\partial\boldsymbol{\Omega}$. According to the Markov Random Field definition, each $\mathbf{a}_i$ is considered as a vertex $\mathcal{V}_i$ and $\mathcal{E}_{ij} = \mathbf{a_i a_j}$ represents a edge between two neighboring nodes *i* and *j*.

Among the traditional exemplar-based methods, when copy a texture patch from the source region Φ to the target region Ω, each Ψ_i have the same orientation as $\hat{\Psi}_{t_i}$, which limits the varieties of the texture synthesis.

Algorithm 3.2 Belief Propagation for Structure Completion

Require: Render each patch Ψ_i along the estimated structures in Ω
Ensure: Find the best matching patches while ensuring texture conherence
1: For each curve $\mathcal{C}$ in Ω, define a series of *anchor points* on it, $\{\mathbf{a}_i, |i = 1 \rightarrow n\}$
2: Collect exemplar-texture patches $\{\hat{\Psi}_{ti}\}$ in Φ, where $t_i \in [1.m]$
3: Setup a factor graph $\mathcal{G} = \{\mathcal{V}, \mathcal{E}\}$ based on $\{\mathcal{C}\}$ and $\{\mathbf{a}_i\}$
4: Defining the energy function $\mathbf{E}$ for each $\mathbf{a}_i$: $\mathbf{E}_i(t_i)$, where t_i is the index in $[1, M]$
5: Defining the message function $\mathbf{M}_{ij}$ for each edge $\mathcal{E}$ in $\mathcal{G}$, with initial value $\mathbf{M}_{ij} \leftarrow 0$
6: Iteratively update all the messages $\mathbf{M}_{ij}$ passed between $\{\mathbf{a_i}\}$
7: $\mathbf{M}_{ij} \leftarrow \min_{ai}\{\mathbf{E}_i(t_i) + E_{ij}(t_i, t_j) + \sum_{k\in N(i), k\neq j} \mathbf{M}_{ki}\}$
8: end until $\Delta\mathbf{M}_{ij} < \delta, \forall i, j$ (by Convergence)
9: Assign the best matching texture patch from $\{\hat{\mathbf{\Psi}}_\mathbf{t}\}$ for each $\mathbf{a}_i$ that $\arg\min_{[T,R]}\{\sum_{i\in\mathcal{V}} \mathbf{E}_i(t_i) + \sum_{(i,j)i\in\mathcal{E}} \mathbf{E}_{ij}(t_i, t_j)\}$, where *T* and *R* represent the translation and orientation of the patch $\{\hat{\mathbf{\Psi}}_{t_i}\}$

Noticing that different patch orientations could produce different results, we introduce a scheme called *Adaptive Patch* by defining a new configuration for the energy metric $\mathbf{E}$ and message $\mathbf{M}$.

Intuitively, the node energy $\mathbf{E}_i(t_i)$can be defined as the *Sum of Square Difference(SSD)* by comparing the known pixels in each patch Ψ_i with the candidate patch in $\hat{\Psi}t_i$. But this could limit the direction changes of the salient structure. So instead of using SSD on the two comparing patches, rotation transformation is performed to the candidate patch before computing the SSD. Mathematically, $\mathbf{E}_i(t_i)$can be formulated as:

$$\mathbf{E}_i(t_i) = \alpha\lambda.P.\sum ||\Psi_i - \dot{R}(\theta).\hat{\Psi}_{ti}||^2_\lambda \tag{6}$$

where $\dot{R}$ represents the 2D rotation matrix with an input angle parameter θ along the orthogonal vector that is perpendicular to the image plane. Since the size of a patch is usually small, the rotation angle θ can be specified with an arbitrary number of values. In our experiment, it is defined as $\theta \in \{0, \pm\frac{\pi}{4}, \pm\frac{\pi}{2}, \pi\}$. Parameter λ represents the number of known pixels in Ψ_i that overlap with the rotated patch $\hat{\Psi}t_i$ P is a penalty term that discourage the candidate patches with smaller proportion of overlapping pixels with the neighboring patches. Here, we define P as $P = \frac{\lambda}{l^2}$ (l is the length of Ψ). α_λ is the corresponding normalization factor.

In a similar way, the energy $E_{ij}(t_i, t_j)$ on each edge $\mathcal{E}_{ij}$can be expressed as:

$$\mathrm{E}_{ij}(t_i, t_j) = \alpha\lambda.\ P.\sum ||\Psi_i(t_i, \theta_{t_i}) - \Psi_j(t_j, \theta_{t_j})||^2_\lambda \tag{7}$$

where i and j *are* the corresponding indices of the two adjacent patches in Ω. The two parameters for Ψ_i indicate the index and rotation for the source patches in $\{\hat{\Psi}_{t_i}\}$. We adopt a similar message passing scheme as [33] that message M_{ij} passes by patches Ψ_i is defined as:

$$M_{ij} = \mathbf{E}_i(t_i) + \mathbf{E}_{ij}(t_i, t_j) \tag{8}$$

Through iterative message passing on the MRF graph to minimize the global energy, an optimal configuration of $\{t_i\}$ for the patches in $\{\Psi_i\}$ can be obtained. The optimal matching patch index $\hat{t_i}$ is defined as:

$$\hat{t_i} = \arg\min_{t_i}\{\mathbf{E}_i(t_i) + \sum_k M_{ki}\} \tag{9}$$

Where k is the index of one of the neighbors of the patch $\Psi_i : k \in \mathcal{N}(i)$. To compute an minimum energy cost, dynamic programming is used: at each step, different states of $\hat{\Psi}_{t_i}$ can be chosen. The edge $\mathcal{E}_{ij}$ represents the transition cost from the state of $\hat{\Psi}_{t_i}$ at step i to state of $\hat{\Psi}_{t_i}$ at step i to state of $\hat{\Psi}t_j$ at step j. Starting from i = 0, an optimal solution is achieved by minimizing the total energy $\xi_i(t_i)$:

$$\xi_i(t_i) = \mathbf{E}_i(t_i) +\ min\{\mathbf{E}_{ij}(t_i, t_j) + \xi_{i-1}(t_{i-1}) \tag{10}$$

where $\xi_i(t_i)$ represents a set of different total energy values at the current step i. In the cases of multiple intersections between curves C, we adopted the idea of Sun's method [33], where readers can refer to for further details.

3.3 Remaining Part Filling

After the structure curves are generated in Ω, we fill the remaining regions by using the exemplar-based approach in [26], where patches are copied from the source region Φ to the filling region Ω in a priority order. The priority is determined by the extracted edges in Φ that intersect with $\partial\boldsymbol{\Omega}$. To ensure the propagated structures in Ω maintain the same orientation as in Φ, higher priorities of texture synthesis are given to those patches that lie on the continuation of stronger edges in Φ.

According to Criminisi's algorithm [26], each pixel on a image has a confidence value and color value. The color value can be empty if it is in the unfilled region Ω. For a given patch Ψ_P at a point **p**, its priority is defined as: *priority* (**p**) = *C*(**p**) . *D*(**p**), where *C*(**p**) and *D*(**p** are the confidence map and data term that are define as:

$$C(p) = \frac{\sum_{P\in\Phi_P \bigcap (\mathcal{I}-\Omega)} \cdot C(\mathbf{q})}{|\Psi_{\mathbf{P}}|} \tag{11}$$

and

$$D(p) = \frac{|\Delta I^{x}_{\mathbf{p}} \,.\, \mathbf{n}^{\perp}_{\mathbf{p}}|}{\alpha} \tag{12}$$

where **q** represents the surrounding pixels of **p** in the patch $\Psi_{\mathbf{P}}$. $|\Psi_{\mathbf{P}}$ is the area of the patch $\Psi_{\mathbf{P}}$. The variable $n_{\mathbf{p}}$ is a unit orthogonal vector that is perpendicular to the boundary $\partial\boldsymbol{\Omega}$ on the point **p**. The normalization factor α is set as 255 as all the pixels are in the range [0, 255] for each color channel. So in such a way, the priority for each pixel can be computed. For further details, we refer readers to the Criminisi's algorithm [26] for more explanations.

4 Experiments

In our experiments, to evaluate our algorithm, different styles of images are tested from natural scenes to indoor environment that has strict structure. Our algorithm obtains satisfactory results in terms of texture coherence and structural consistency. The algorithm is implemented in C++ code with OpenCV library. All the images results are generated on a dual-core PC with

CPU 2.13GHz and Memory 2G. For the images with the regular resolution 640 x 480, the average computation cost is about 52 seconds.

To verify the performance of the algorithm, we first compare the result of our method with the one proposed in [26] on the well-known *Kanizsa triangle* in terms of structure coherence and texture consistency. As shown in Figure 5(a), the white triangle in the front is considered as Ω that needs to be filled. First, a structure propagation is performed based on the detected edges along $\partial\Omega$. The dash curves in Figure 5(b) indicate the estimated potential structure in the missing area Ω, which are generated by our structure propagation algorithm. Texture propagation is applied to the rest of the image based on the confidence and isophote terms. One can notice both the triangle

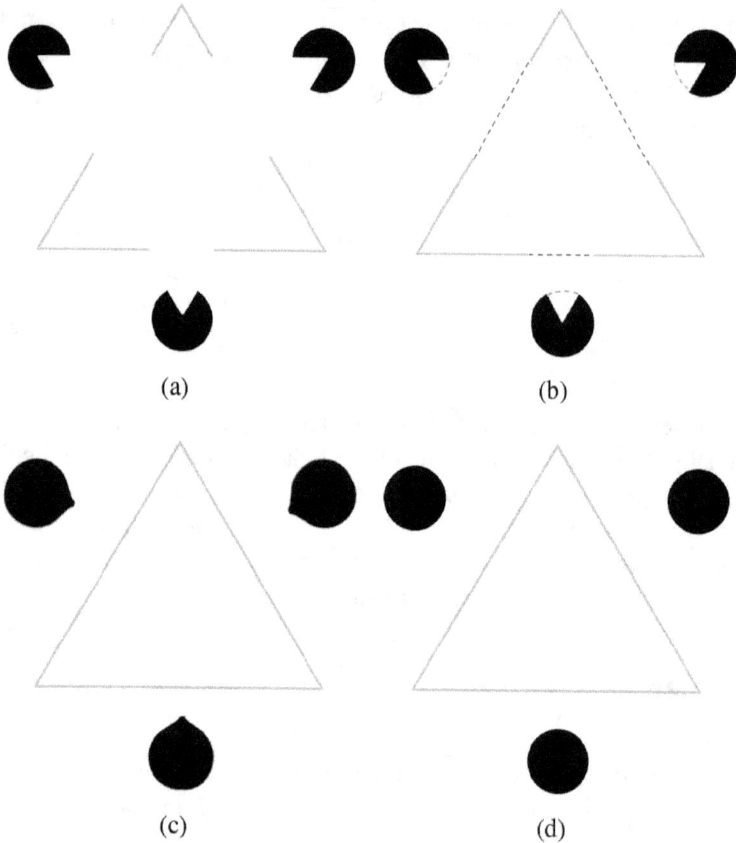

Figure 5 Kanizsa triangle experiment (a) Original image (b) Curve reconstruction for the missing region Ω (c) Result image by criminisi's method (d) Result image by our method

and the circles are well completed in our result Figure 5(d) comparing with Criminisi's method in Figure 5(c).

Figure 6 demonstrates the advantage of our method by preserving the scene structure after removing the occluded object. The original image data can be find publicly at the website [1]. Figure 6(b) shows the target region (the bungee jumper) for removal marked in green color. Figure 6(c), 6(d) are the image reconstruction results by the Criminisi's and our methods respectively. One can notice the roof area in Figure 6(c) is broken by the grass which introduces noticeable artifact, while the corresponding part remains intact in our result. Furthermore, in contrast to the Criminisi's method, the lake boundary is naturally recovered thanks to our structure estimation procedure,

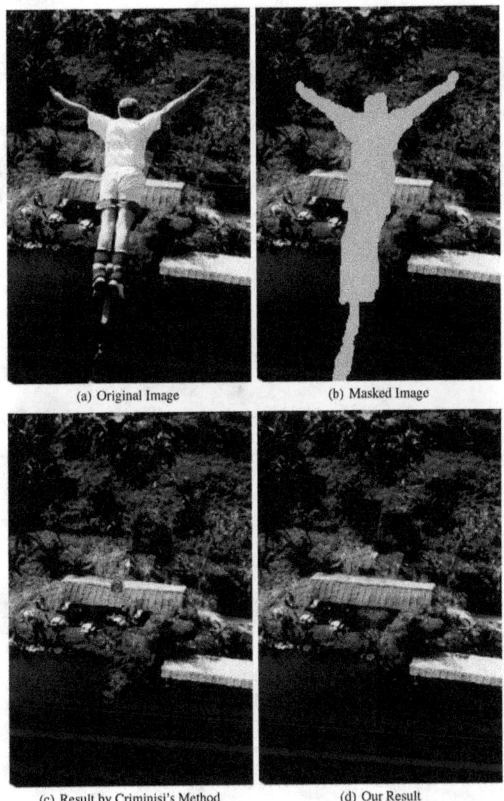
(a) Original Image (b) Masked Image

(c) Result by Criminisi's Method (d) Our Result

Figure 6 Result comparison with criminisi's method [26]

[1] http://www.cc.gatech.edu/sooraj/inpainting/

as shown in 6(d). In terms of the time performance for the original image of 205 × 307 pixels, our method performed 10.5 seconds on the computer of dual-core PC with CPU 2.13GHz and 2GB of RAM, to be compared with 18 seconds of Criminisi's on a 2.5 GHz Pentium IV with 1 GB of RAM.

Another existing work we choose to compare with is the Sun's method, which also aims to preserve the original structure in the recovered image. However, the difference is that Sun's method requires manual intervention during the completion process. The potential structure in the target region needs to be manually labeled by the designer, which can vary according to individuals. Figure 7 demonstrates a comparison between Sun's and our methods. In the original image, the car is considered as the target object for removal, which is marked in black color in Figure 7(b). In Figures 7(c), 7(d),

Figure 7 Result comparison with sun's method [33]

the potential structures in the target region are labeled by [33] and automatically estimated by our method, which produce different results, as shown in Figures 7(e), 7(f). To compare the computation speed, our methods performed 51.7 seconds to process this image (640 × 457), in contrast with the Sun's fewer than 3 seconds for each curve propagation (3 curves in total) and 2 to 20 seconds for each subregion (4 subregions in total) on a 2.8 GHz PC. Moreover, we save the potential labor work on specifying the missing structures by the user.

To further demonstrate the performance, a set of images are used for scene recovery: ranging from indoor environment to natural scenes. Figure 8 shows the case of indoor environment, where highly structural patterns often present, such as the furniture, windows, walls. In Figure 8, the green bottle on the office partition is successfully removed while preserving the remaining structure. In this example, five pairs of edges are identified and connected by the corresponding curves that are generated in the occluded region Ω. Guided by the estimated structure, plausible texture information is synthesized to form a smooth intensity transition across the occluded region with little artifact.

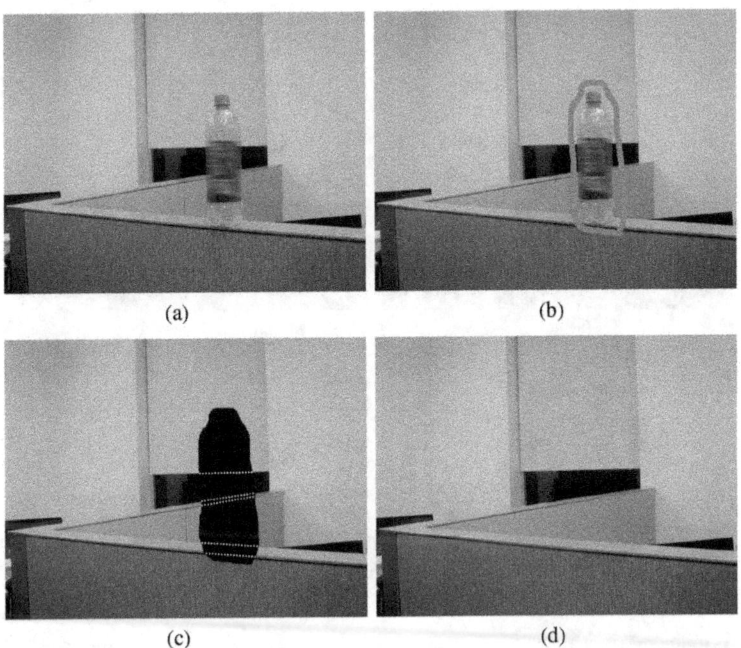

Figure 8 Demo 1- Bottle removal (a) Original image (b) Image with user's label for removal (c) Generated structure in the missing region (d) Result image

Figure 9 Demo 2- Tree removal 1 (a) Original image (b) Image with user's label for removal (c) Generated structure in the missing region (d) Result image

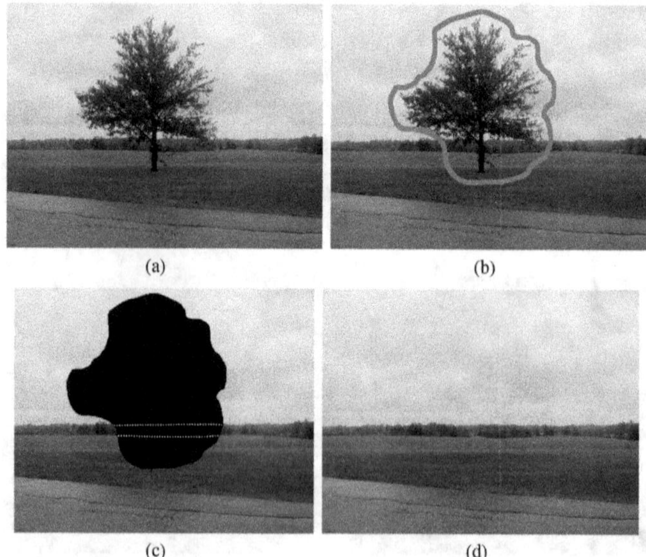

Figure 10 Demo 3- Tree removal 2 (a) Original image (b) Image with user's label for removal (c) Generated structure in the missing region (d) Result image

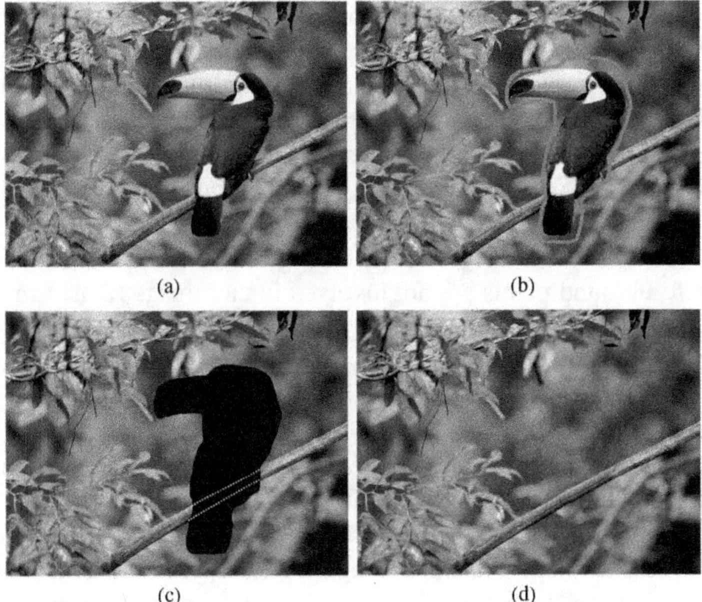

Figure 11 Demo 4- Bird removal (a) Original image (b) Image with user's label for removal (c) Generated structure in the missing region (d) Result image

For the outdoor environment, as there are fewer straight lines or repeating patterns existing in the natural world, the algorithm should provide the flexibility to generate irregular structures. Figure 9 and 10 show the results of removing trees in the nature scenes. Several curves are inferred by matching the broken edges along $\partial\Omega$ and maximizing the continuity. We can notice the three layers of the scene (sky, background trees, and grass land) are well completed. In Figure 11, it shows a case that a perching bird is removed from the tree. Our structure estimation successfully completes the tree branch with smooth geometric and texture transitions. Without such structure guidance, those traditional exemplar-texture based methods can easily produce noticeable artifacts. For example, multiple tree branches may be generated as the in-painting process and directions largely rely on the matching patches.

5 Conclusion

In this paper, we present a novel approach for foreground objects removal while ensure structure coherence and texture consistency. The core of our approach is to use structure as a guidance to complete the remaining scene, which demonstrates its accuracy and consistency. This work would benefit a wide range of applications, from digital image restoration (e.g. scratch recovery) to privacy protection (e.g. remove people from the scene). In particular, this technique can be promising for the online massive collections of imagery, such as photo localization and scene reconstructions. By removing foreground objects, the matching accuracy can be dramatically improved as the corresponding features are only extracted from the static scene rather than those moving objects. Furthermore it can generate more realistic views because the foreground pixels are not involved in any image transformation and geometric estimation.

As one direction of our future work, we will apply this object removal technique to scene reconstruction applications that can generate virtual views or reconstruct the 3D data from a set of images. Multiple images can give more cues of the potential structure and texture in the target region. For example, through corresponding features among different images, intrinsic and extrinsic parameters can be estimated. Then the structure and texture information can be mapped from one image to another image. So for a particular target region for completion, multiple sources (from different images) can contribute the estimation. As such, Our current algorithm needs to be modified adaptively to take the advantage of the extra information. An optimization framework could be established to identify optimal structures and textures to fill the target region.

References

[1] Jim Jing-Yan Wang, Halima Bensmail, and Xin Gao. Joint learning and weighting of visual vocabulary for bag-of-feature based tissue classification. *Pattern Recognition*, 46(12):3249–3255, 2013.

[2] Jim Jing-Yan Wang, Halima Bensmail, and Xin Gao. Multiple graph regularized nonnegative matrix factorization. *Pattern Recognition*, 46(10):2840–2847, 2013.

[3] Q. Sun, P. Wu, and Y. Wu. Unsupervised multi-level non-negative matrix factorization model: Binarydata case. *International Journal of Information Security*, 3(4):245–250, 2012.

[4] Jing-Yan Wang, Islam Almasri, and Xin Gao. Adaptive graph regularized nonnegative matrix factorization via feature selection. In *Pattern Recognition (ICPR), 2012 21st International Conference on*, pages 963–966. IEEE, 2012.

[5] Ju Shen and S.-C.S. Cheung. Layer depth denoising and completion for structured-light rgb-d cameras. In *Computer Vision and Pattern Recognition (CVPR), 2013 IEEE Conference on*, pages 1187–1194, 2013.

[6] Yuzhu Zhou, Le Li, Tong Zhao, and Honggang Zhang. Region-based high-level semantics extraction with cedd. In *Network Infrastructure and Digital Content, 2010 2nd IEEE International Conference on*, pages 404–408. IEEE, 2010.

[7] Ju Shen and Wai tian Tan. Image-based indoor place-finder using image to plane matching. In *Multimedia and Expo (ICME), 2013 IEEE International Conference on*, pages 1–6, 2013.

[8] Yuzhu Zhou, Le Li, and Honggang Zhang. Adaptive learning of region-based plsa model for total scene annotation. *arXiv preprint arXiv:1311.5590*, 2013.

[9] Le Li, Jianjun Yang, Kaili Zhao, Yang Xu, Honggang Zhang, and Zhuoyi Fan. Graph regularized non-negative matrix factorization by maximizing correntropy. *arXiv preprint arXiv:1405.2246*, 2014.

[10] Q. Sun, F. Hu, and Q. Hao. Context awareness emergence for distributed binary pyroelectric sensors. In *Proceeding of 2010 IEEE International Conference of Multi-sensor Fusion and Integration of Intelligent Systems*, pages 162–167, 2010.

[11] Le Li, Jianjun Yang, Yang Xu, Zhen Qin, and Honggang Zhang. Document clustering based on max-correntropy non-negative matrix factorization. 2014.

[12] Q. Sun, F. Hu, and Q. Hao. Mobile target scenario recognition via low-cost pyroelectric sensing system: Toward a context-enhanced accurate identification. *IEEE Transactions on Systems, Man and Cybernetics: Systems*, 4(3):375–384, 2014.

[13] Ju Shen, Anusha Raghunathan, Sen ching Cheung, and Rita Patel. Automatic content generation for video self modeling. In *Multimedia and Expo (ICME), 2011 IEEE International Conference on*, pages 1–6, 2011.

[14] Ju Shen, Changpeng Ti, Sen ching Cheung, and R.R. Patel. Automatic lip-synchronized video-self-modeling intervention for voice disorders. In *e-Health Networking, Applications and Services (Healthcom), 2012 IEEE 14th International Conference on*, pages 244–249, 2012.

[15] Ju Shen, Changpeng Ti, Anusha Raghunathan, Sen ching Cheung, and Rita Patel. Automatic video self modeling for voice disorder. In *Computer Vision and Pattern Recognition (CVPR), 2013 IEEE Conference on*, pages 1187–1194, 2013.

[16] Li Xu, Zhenxin Zhan, Shouhuai Xu, and Keying Ye. Cross-layer detection of malicious websites. In *Proceedings of the third ACM conference on Data and application security and privacy*, pages 141–152. ACM, 2013.
[17] Jianjun Yang and Zongming Fei. Hdar: Hole detection and adaptive geographic routing for ad hoc networks. In *Computer Communications and Networks (ICCCN), 2010 Proceedings of 19th International Conference on*, August 2010.
[18] Li Xu, Zhenxin Zhan, Shouhuai Xu, and Keying Ye. An evasion and Counter-Evasion study in malicious websites detection. In *2014 IEEE Conference on Communications and Network Security (CNS) (IEEE CNS 2014)*, San Francisco, USA, 2014.
[19] Jianjun Yang and Zongming Fei. Broadcasting with prediction and selective forwarding in vehicular networks. In *International Journal of Distributed Sensor Networks*, 2013.
[20] Yi Wang, Arnab Nandi, and Gagan Agrawal. SAGA: Array Storage as a DB with Support for Structural Aggregations. In *Proceedings of SSDBM*, June 2014.
[21] Hong Zhang, Fengchong Kong, Lei Ren, and Jian-Yue Jin. An interprojection interpolation (ipi) approach with geometric model restriction to reduce image dose in cone beam ct (cbct) computational modeling of objects presented in images. In *Fundamentals, Methods, and Applications Lecture Notes in Computer Science*, volume 8641, pages 12–23, 2014.
[22] Yi Wang, Wei Jiang, and Gagan Agrawal. Scimate: A novel mapreduce-like framework for multiple scientific data formats. In *Cluster, Cloud and Grid Computing (CCGrid), 2012 12th IEEE/ACM International Symposium on*, pages 443–450. IEEE, 2012.
[23] Hong Zhang, Vitaly Kheyfets, and Ender Finol. Robust infrarenal aortic aneurysm lumen centerline detection for rupture status classication. In *Medical Engineering and Physics*, volume 35(9), pages 1358–1367, 2013.
[24] Hong Zhang, Lin Yang, David J. Foran, John L. Nosher, and Peter J. Yim. 3d segmentation of the liver using free-form deformation based on boosting and deformation gradients. In *IEEE International Symposium on Biomedical Imaging (ISBI)*, 2009.
[25] Bertalmio M. Sapiro G. Ballester C. and Caselles V. Image inpainting. *Proceedings of ACM SIGGRAPH 2000*, pages 417–424, 2000.

[26] A. Criminisi P. Prez and K. Toyama. Region filling and object removal by exemplar-based inpainting. *IEEE Trans. Image Process*, 13(9): 1200–1212, Sep 2004.

[27] M. Ashikhmin. Synthesizing natural textures. *Proc. ACM Symposium on Interactive 3D Graphics*, pages 217–226, March 2001.

[28] A. Efros and W.T. Freeman. Image quilting for texture synthesis and transfer. *Proc. ACM Conf. Comp. Graphics (SIGGRAPH)*, pages 341–346, August 2001.

[29] A. Efros and T. Leung. Texture synthesis by non-parametric sampling. *Proc. Int. Conf. Computer Vision*, pages 1033–1038, September 1999.

[30] M. Bertalmio L. Vese G. Sapiro and S. Osher. Simultaneous structure and texture image inpainting. *Proc. Conf. Comp. Vision Pattern Rec.*, 2003.

[31] I. Drori D. Cohen-Or and H. Yeshurun. Fragment-based image completion. *ACM Trans. on Graphics*, 22(3):303–312, 2003.

[32] J. Jia and C.-K. Tang. Image repairing: Robust image synthesis by adaptive nd tensor voting. *Proc. Conf. Comp. Vision Pattern Rec.*, 2003.

[33] Jian Sun. Lu Yuan. Jiaya Jia. and Heung-Yeung Shum. Image completion with structure propagation. *ACM Transactions on Graphics (TOG) - Proceedings of ACM SIGGRAPH*, 2005.

[34] Jianjun Yang, Yin Wang, Honggang Wang, Kun Hua, Wei Wang, and Ju Shen. Automatica objects removal for scene completion. *IEEE INFOCOM Workshop on Security and Privacy in Big Data, Toronto, Canada*, 2014.

[35] Pablo Arbeláez. Michael Maire. Charless Fowlkes and Jitendra Malik. Contour detection and hierarchical image segmentation. IEEE *Transactions on Pattern Analysis and Machine Intelligence*, 33:898–916, May 2011.

[36] Lab. d'Anal. Numerique. Univ. Pierre et Marie Curie. Paris. Disocclusion: a variational approach using level lines. *IEEE Transactions on Image Processing*, 11:68–76, Feb 2002.

[37] Pablo Arbeláez. Boundary extraction in natural images using ultrametric contour maps. *Conference on Computer Vision and Pattern Recognition Workshop, 2006. CVPRW '06*, pages 182–190, Jun 2006.

[38] P. W. Lamberti, A. P. Majtey, A. Borras, M. Casas, and A. Plastino. Metric character of the quantum jensen-shannon divergence. *Phys. Rev. A*, 77:052311, May 2008.

39 James McCrae and Karan Singh. Sketching piecewise clothoid curves. *Computers and Graphics*, 33:452–461, Aug 2009.

[39] Mullinex G and Robinson ST. Fairing point sets using curvature. *Computer Aided Design*, 39:27–34, 2007.

Biographies

Ju Shen is an Assistant Professor from the Department of Computer Science, University of Dayton (UD), Dayton, Ohio, USA. He received his Ph.D. degree from University of Kentucky, Lexington, KY, in USA and his M.Sc degree from University of Birmingham, Birmingham, United Kingdom. His work spans a number of different areas in computer vision, graphics, multimedia, and image processing.

Jianjun Yang Jianjun Yang received his B.S. degree and his first M.S. degree in Computer Science in China, his second M.S. degree in Computer Science during his doctoral study from University of Kentucky, USA in May 2009, and his Ph.D degree in Computer Science from University of Kentucky, USA in 2011. He is currently an Assistant Professor in the Department of Computer Science and Information Systems at the University of North Georgia. His research interest includes wireless networking, computer networks, and image processing.

Sami Taha Abu Sneineh is currently an assistant professor at Palestine Polytechnic University (PPU). He received his Ph.D. degree in May 2013 in Computer Vision at the University of Kentucky (UK) under the supervision of Dr. Brent Seales. Sami earned a B.Sc in Electrical Engineering from Palestine Polytechnic University in 2001 and M.Sc in Computer Science from Maharishi University of Management in 2006. His research focus is on computer vision in minimally invasive surgery to improve the performance assessment. Sami has worked for three years as consultant software engineer at Lexmark Inc. and four years as consultant programmer at IBM before joining UK. He started his teaching experience as TA in the computer science department. He received a certificate and an award in college teaching and learning in 2012. His goal is to contribute in improving the higher education system and raise the education standards.

Bryson Payne received his B.S. degree in Mathematics from North Georgia College & State University, USA in 1997, his M.S. degree Mathematics from North Georgia College & State University, USA in 1999, and his Ph.D degree in computer science from Georgia State University, USA in 2004. He was the (CIO) Chief Information Officer in the North Georgia College & State University (former University of North Georgia). He is currently an Associate Professor in the Department of Computer Science and Information Systems at the University of North Georgia. He is also a CISSP (Certified Information Systems Security Professional). His research interest includes image processing, web application and communications.

Markus Hitz received his B.S. degree and M.S. degree of Computer Science in Europe and his Ph.D degree of Computer Science in USA. He is currently a Professor and Acting Head in the Department of Computer Science and Information Systems at the University of North Georgia. His research interest includes networking, simulation and communications.

Evaluation and Analysis of Distributed Graph-Parallel Processing Frameworks

Yue Zhao,[1] Kenji Yoshigoe,[1] Mengjun Xie,[1] Suijian Zhou,[1] Remzi Seker[2] and Jiang Bian[3*]

[1]*Department of Computer Science, University of Arkansas at Little Rock, Little Rock, AR 72204, USA*
[2]*Department of ECSSE, Embry-Riddle Aeronautical University, Daytona Beach, FL 32114, USA*
[3]*Department of Biomedical Informatics, University of Arkansas for Medical Sciences, Little Rock, AR 72205, USA*
*{yxzhao; kxyoshigoe; mxxie; sxzhou}@ualr.edu, sekerr@erau.edu, jbian@uams.edu (*Corresponding)*

Received 15 June 2014; Accepted 20 August 2014
Publication 7 October 2014

Abstract

A number of graph-parallel processing frameworks have been proposed to address the needs of processing complex and large-scale graph structured datasets in recent years. Although significant performance improvement made by those frameworks were reported, comparative advantages of each of these frameworks over the others have not been fully studied, which impedes the best utilization of those frameworks for a specific graph computing task and setting. In this work, we conducted a comparison study on parallel processing systems for large-scale graph computations in a systematic manner, aiming to reveal the characteristics of those systems in performing common graph algorithms with real-world datasets on the same ground. We selected three popular graph-parallel processing frameworks (Giraph, GPS and GraphLab) for the study and also include a representative general data-parallel computing system— Spark—in the comparison in order to understand how well a general data-parallel system can run graph problems. We applied basic performance

Journal of Cyber Security, Vol. 3, 289–316.
doi: 10.13052/jcsm2245-1439.333

metrics measuring speed, resource utilization, and scalability to answer a basic question of which graph-parallel processing platform is better suited for what applications and datasets. Three widely-used graph algorithms—clustering coefficient, shortest path length, and PageRank score—were used for benchmarking on the targeted computing systems. We ran those algorithms against three real world network datasets with diverse characteristics and scales on a research cluster and have obtained a number of interesting observations. For instance, all evaluated systems showed poor scalability (i.e., the runtime increases with more computing nodes) with small datasets likely due to communication overhead. Further, out of the evaluated graph-parallel computing platforms, PowerGraph consistently exhibits better performance than others.

Keywords: Big data, Graph-parallel computing, Distributed processing.

1 Introduction

Recent years have seen the exponential growth of scale and complexity of networks from various disciplines, sectors, and applications such as World Wide Web, social networks, brain neural networks, transportation networks and so on. These real-world networks are often modeled as graphs and studied by applying graph theories. It becomes increasingly important to gain insights and discover knowledge from those large, real-world networks, e.g., identifying critical nodes in information distribution by studying social networks and discovering biomarkers from gene regulatory networks and human brain connectome [13]. However, gigantic size of those graphs that consist of millions (or even billions) of vertices and hundreds of millions (or billions) of edges poses a great challenge to study them as effective parallelization of graph computations becomes the key.

There are many parallel computing paradigms—e.g., Message Passing Interface (MPI) [1], Open MultiProcessing (OpenMP) [2], MapReduce [3], and graph-parallel computing systems [4–6]—available for parallel processing. The general idea of parallelizing computational and/or data intensive tasks is to split a large computing job into multiple smaller tasks and distribute them onto multiple computing machines for parallel processing.

Traditionally, MPI has been very popular as it provides essential virtual topology, synchronization, and communication functionality between a set of processes (that have been mapped to computer nodes) in a language-independent way. MPI achieves these goals through standardizing message passing between processes on parallel computers. On the other hand, OpenMP provides shared memory parallelism with a set of compiler directives and library routines for Fortran and C/C++ programs. The learning curve for programming in MPI and OpenMP environments has been quite steep even after many helper software packages (e.g., PETSc [7]) have become available to ease the use of MPI. On the other hand, MapReduce [3] is significantly easier to learn because of a well-defined programming model. In MapReduce, nodes communicate through disk I/O (i.e., a shared distributed file system) while nodes in MPI exchanges data and states by message passing. Consequently, in general, MapReduce is more suitable for data-intensive tasks (data-parallelism) where nodes require little data exchange to proceed while MPI is more appropriate for computation-intensive tasks (task-parallelism) [8]. Moreover, the support of fault tolerance is a built-in feature in MapReduce, making application development more user-friendly, transparent and easy to debug.

Unfortunately, traditional data-parallel computing systems such as MapReduce [3] and Spark [9] cannot take advantage of the characteristics of graph-structure data and often result in complex job chains and excessive data movement when implementing iterative graph algorithms. In order to seek the leap of performance on processing graph data, numerous specialized graph-computing engines such as Pregel [14], and Graphlab [4–6] have been developed.

Effective parallelization of computing tasks can vary substantially depending on the characteristics of computation (e.g., task parallelism vs. data parallelism and fine-grained parallelism vs. coarse-grained parallelism vs. embarrassing parallelism). For example, MapReduce based frameworks are extremely good at solving SIMD (single-instruction, multiple-data) problems. Efficient parallelism arises by breaking a large dataset into independent parts with no forward or backward dependencies in each Map-Reduce pair. In contrast, parallelizing machine learning algorithms such as logistic regression and random forest requires a different computing model, as the algorithms are often iterative where future iterations have a high-level of data dependency over previous results. Running iterative algorithms using MapReduce is shown to incur excessive communications and hurt the performance [9]. A comprehensive understanding of the characteristics of different big data computing platforms through quantitative and qualitative measurements is necessary to understand their respective strengths in handling different types of computational tasks.

In this paper, we studied different distributed graph-parallel computing systems processing graph-structured data due to: 1) the growing importance and popularity of graph data in both industry and academia; and 2) better performance over data-parallel platforms reported in existing work for certain computation tasks [15, 16]. Our analyses aim to not only evaluate and compare the basic performance metrics (e.g., runtime, resource utilization, and scalability) of these systems, but also to present instructive experience for selecting the most appropriate data processing platforms based on characteristics of the applications and/or datasets. Spark [9, 10] was included in this study, as a representative of data-parallel processing systems, with the intent to better understand the difference between graph-parallel and data-parallel platforms. Our evaluation was conducted by executing three important graph-processing algorithms (i.e., PageRank, clustering coefficient and shortest path length) under the targeted systems (i.e., Spark [9, 10], Graphlab [4–6], GPS [17], Pregel/Giraph [14, 18]). The input data for the experiments are three graph-structured datasets extracted from real-world use with diverse characteristics and different scales in size.

The experiment results show that 1) comparing with Spark - a data-parallel processing system, graph-parallel computing platforms exhibited better performance in terms of graph computing rate and resource utilization; 2) PowerGraph outperformed others graph-parallel computing platforms under most evaluation cases; and 3) all evaluated graph-parallel computing systems exhibited different scalability on datasets with diverse scales in size.

The rest of the paper is organized as follows. Section 2 describes our design for evaluating and analyzing distributed graph-parallel computing platforms. Experiment results and analysis are presented in Section 3. Section 4 overviews related work, and Section 5 concludes this paper.

2 Design of experiments

In this section we detail the design of the experiments for evaluating distributed graph-parallel processing platforms. Our experiments focused on graph algorithms and graph-structured datasets as they were aimed to examine whether and how graph-parallel systems improve the efficiency of executing large-scale graph algorithms compared to general-purpose data-parallel systems. The experiment design essentially had three key considerations: 1) identifying performance metrics that system users are mostly concerned with; 2) selecting representative graph algorithms and real-world datasets of different scale; and 3) identifying representative graph-parallel computing systems. We conducted experiments on each of the selected parallel computing system, compared and analyzed the results in terms of both the quantitative performance metrics and qualitative user experiences of the systems.

2.1 Performance Metrics

The performance metrics we considered in the experiments are mainly from the perspective of an end user of a parallel computing platform. Nowadays big data computation jobs including those large-scale graph computations are quite likely to be executed in a computing cloud environment leased by the user. Obviously, the sooner the job completes and the less the resources are used, the better. Therefore, we chose metrics that measure data processing rate (speed) and resource utilization. We also studied the system scalability which is directly related to the processing rate.

1) *Data Processing Rate:* Data processing rate measures how quickly a distributed computing platform can process data and finish the execution of the computation job. We are interested in the overall runtime spent by the system on processing a particular dataset. The entire runtime consists of two parts: the data ingress time and job execution time. The former refers to the amount of time spent by the system from bootstrapping to the completion of data ingestion and the latter refers to the amount of time spent by the system in executing the computation job. Typically, in a distributed environment, the ingress time consists of the time for bootstrapping the computing cloud (or cluster) and the time for reading and partitioning the data and the ingress time is usually dominated by the latter.

2) *Scalability:* Scalability is the ability of a computing system to accommodate the growth of the amount of work by adding more computing resources (e.g., compute nodes) without changing the system itself. For distributed computing systems, a larger dataset usually indicates more computations. We studied the scalability of the selected distributed systems by examining: 1) the execution time of the same computation job with different size of dataset; and 2) the speedup achieved by using more compute nodes when dealing with the same dataset.

3) *Resource Utilization:* Resource utilization measures the degree of usage for each type of hardware and software resources (e.g., CPU, memory, file system cache, etc), which helps to understand the running behavior and efficiency of a computing platform. For a distributed computing system, the most concerned

computing resources are CPU, memory, and network bandwidth. Thus, we selected the following metrics: CPU load, amount of consumed memory, and network I/O volume, to study the resource utilization of the targeted platforms.

In summary, the performance metrics studied in this work include data ingress time, job execution time, CPU load, memory consumption, and network I/O volume. Since the platforms we studied are distributed computing systems, the CPU load, memory consumption and network I/O volume are normalized by the number of computing nodes if not specified.

2.2 Benchmarking Datasets and Graph Algorithms

This section discusses the selected benchmarking datasets and graphs algorithms.

1) *Benchmarking Datasets*

We selected three graph datasets extracted from real-world social networks with diverse characteristics (e.g. graph density, average vertex degree (AVD), and directivity) and different scales (in terms of number of vertices and edges). We included both the directed and undirected graphs. Table 1 summarizes the datasets used in this study. Dataset G1, a sampled snapshot of the Facebook social graph obtained in 2009 [23], has the largest number of vertices and edges. Datasets G2 and G3 reflect LiveJournal friendship network (directed) and DBLP coauthorship network (undirected), respectively. Both of them are obtained from Stanford Network Analysis Project (SNAP)[1]. The graph density measures the proportion of the number of actual edges over the maximal possible number of edges. It is defined as $2|E|/(|V|(|V|-1))$ for undirected graph and $|E|/(|V|(|V|-1))$ for directed graph. The average vertex degree is defined as $|E|/|V|$ for directed graph and $2|E|/|V|$ for undirected graph.

2) *Graph Algorithms*

Table 2 lists the three graph algorithms being studied. We chose those algorithms because they are basic graph algorithms and commonly used in network analysis studies. Those algorithms often are already implemented in distributed graph-parallel computing systems and used as the benchmarking algorithms to evaluate system performance in existing studies (e.g. [4][6]).

[1] http://snap.stanford.edu/data/

Table 1 Summary of the experiment datasets

Dataset	Graphs	Description	# of vertices (|V|)	# of edges (|E|)	Graph density ($\times 10^{-5}$)	AVD	Directivity
G1	BFS1	Facebook social network	61,876,615	336,776,269	0.02	5	Directed
G2	Soc-LiveJournal	LiveJournal friednship social network	4,847,571	68,993,773	0.59	14	Directed
G3	Com-DBLP	DBLP co-authorship network	317,080	1,049,866	2.09	7	Undirected

Table 2 Summary of the studied graph algorithms

ID	Algorithm	Characteristics	Example Application
A1	PageRank	Iterative, high communication	Importance ranking
A2	Shortest path	Iterative, low communication	Decision making
A3	Triangle counting	Single step, medium communication	Clustering coefficient

PageRank [19] and its variants such as Personalized PageRank [20] are effective methods for link prediction based on finding structure similarities between nodes in a network. Conceptually, the PageRank score of a node is the long-term probability that a random web surfer is at that node at a particular time step. For sufficiently long time, the probability distribution of the random walks on the graph is unique, that is, minor changes to the graph make the random walk transition matrix aperiodic and irreducible [21]. The computation of the PageRank score of a node *v* is an iterative process where the PageRank algorithm recursively computes the rank *R(v)* considering the scores of nodes *u* that are connected to *v*, defined as:

$$R(v) = (1 - \alpha) \sum_{u\ links\ to\ v} w_{u,v} \times R(u) + \frac{\alpha}{n}$$

where α is the damping factor (see Langville et. al.'s "Deeper Inside PageRank" for a detailed and excellent description of the PageRank algorithm).

Topological features of a network can be quantitatively measured as network characteristics such as the clustering coefficient and characteristic path length. These structural network characteristics are often used to benchmark or infer the functional aspects of the network.

The clustering coefficient of a vertex expresses the chance of how likely its neighbors are also connected to one another. The (global) clustering coefficient is based on triplets of nodes, where a triplet consists of three nodes that are connected by either two (open) or three (closed) undirected ties. The global clustering coefficient is therefore defined as the fraction of the number of closed triplets over number of total connected triplets of vertices. Therefore, the method of calculating the global clustering coefficient is also often called triangle counting [22]. The clustering coefficient measures the degree of herding effect in a network (or network component), a larger coefficient value implies that nodes tend to create more tightly knit groups.

The characteristic path length is the average shortest path length in a network [22], which measures the average degree of separation between nodes in a network (or network component). Therefore, the shorter the length, the "easier" (or more likely) it can reach another node in a network.

The clustering coefficient and the characteristic path length are important network measures that are often used to determine the type of a network (e.g., random, small-world and scale-free). For example, Watts and Strogatz [22] coined the term "small-world" networks to categorize complex sparse real-life networks that have significantly high clustering coefficients than sparse random graphs yet have small degrees of separation (i.e., characteristic path length) between nodes.

2.3 Distributed Big Data Platforms and the Targeted Platforms

As this work focuses on the evaluation of graph-parallel processing platforms, we selected three systems that are recent and popular both in academia and industry. They are Giraph, GPS, and PowerGraph (the upgraded version of GraphLab). In addition, we also included a popular general-purpose data-parallel computing platform, Spark, into the study in order to understand the performance difference between a data-parallel computing platform and graph-parallel computing platform in executing a graph computation job. The general information of the computing platforms evaluated in this study is summarized in Table 3. Note that Spark, Giraph, and GPS use only synchronous computation mode while both synchronous and asynchronous modes are supported by PowerGraph and were tested in our experiments.

1) Data-Parallel Computing System

Data-parallel computing systems use a simple programming model to process large-scale data. Whereby the data-parallelism feature, these systems support and implement a couple of fault-tolerance strategies and are highly scalable [24]. They are able to process data organized in any kinds of format, such as table and graph. Thus, data-parallel computing systems are also tagged as generic data processing systems [11]. MapReduce/Hadoop [3] and Spark [9] are the most popular representatives of data-parallel computing

Table 3 Summary of the studied distributed computing platforms

Platform	Type	Version	Release Year	Computation model
Spark	Data parallel	Spark-0.9.0	2014	Synchronous
PowerGraph	Graph parallel	PowerGraph-2.2	2012	GAS (Synchronous & Asynchronous)
Giraph	Graph parallel	Giraph 1.0.0	2013	BPS (Synchronous)
GPS	Graph parallel	GPS-0.0.1	2013	BPS (Synchronous)

systems. Because literatures [9, 27] have demonstrated that Spark Mapreduce: simplified data processing on large clusters performs much better than Mapreduce/Hadoop in the big data processing rate and efficiency, we selected Spark, but not Mapreduce/Hadoop, as a representative of data-parallel computing systems to evaluate and analyze in this work.

Spark [9]: While retaining the fault tolerance and scalability of MapReduce Spark proposes three simple data abstractions: two restricted types of shared variables (broadcast variables and accumulators) and the resilient distributed datasets (RDDs). In particular, RDDs are read-only partitioned objects stored separately on a set of computing nodes. It can be rebuilt if a partition is lost. Spark can outperform Hadoop by 10x in executing time for iterative machine learning jobs. The graph analytic library, GraphX [24], on Spark efficiently formulates graph computation within the Spark data-parallel framework, and it exploits the special graph data structures such as vertex cuts, structural indices and graph-parallel operators.

2) Graph-Parallel Computing System

Graph-parallel platforms are designed and developed specially for processing graph data. In a graph-parallel computing system, the input of an algorithm is presented as a sparse graph $G = \{V, E\}$, and the computation is conducted by executing a vertex-program Q in parallel on each vertex $v \in V$. The vertex-program $Q(v)$ can communicate with neighboring instances $Q(u)$ where $(u, v) \in E$. For example, GraphLab [5] provides a shared data table (SDT), an associative map, to support globally shared information and Pregel [14] adopts message-passing model to exchange information among vertex-program instances [3]. In the following paragraphs we introduce the targeted graph-parallel computing systems evaluated in this work and demonstrate their characteristics to clarify the reason why we selected them.

PowerGraph [6]: GraphLab exhibits more competitive performance than others in graph processing [11, 15, 24]. Besides bulk-synchronous computation model GraphLab also supports asynchronous graph computation to alleviate the overhead induced by waiting for barrier synchronization. PowerGraph is a representative of GraphLab platform series [4, 5, 6, 28] and it is an open-source, distributed graph-specific computing system implemented in C++. Based on the primary distributed Graphlab [4] PowerGraph makes the following improvements. First, PowerGraph proposes a three-phase programming model, Gather, Apply and Scatter (GAS), for constructing a vertex-program. By directly exploiting the GAS decomposition to factor vertex programs over edges, PowerGraph eliminates the degree dependence of the vertex-program [6]. Second, PowerGraph incorporates the best features from both Pregel [14] and GraphLab [4]. Like GraphLab, PowerGraph adopts the shared-memory and data-graph view of computation that frees users from architecting a communication protocol for sharing information. PowerGraph borrows the commutative associative message combiner from Pregel, which reduces communication overhead in the Gather phase. At last, PowerGraph uses a vertex-cut approach to address the issue of partitioning power-law graphs [6]. Vertex-cutting can quickly partition a large power-law graph by cutting a small fraction of very high degree vertices.

Giraph [18]: Pregel [14] is a most popular and representative bulk synchronous parallel (BSP) computing system, in which all vertex-programs run simultaneously in a series of global super-steps. Many other state-of-the-art graph-parallel systems [17, 18, 29] derive from Pregel. Pregel employs message passing model to exchange information among vertex-computing instances. However, the source code of Pregel is not open and it is not feasible to evaluate it directly, thus we selected its public implementation: Giraph to evaluate in this work. Giraph inherits the vertex-centric programming model of Pregel and isan open-source, distributed graph-parallel processing platform. Giraph leverages the Map phase from Mapreduce/Hadoop and to achieve fault tolerance Giraph adopts the periodic checkpoints. Like Graphlab, Giraph is also executed in-memory, in which a whole graph needs to be imported into memory to process. Although this feature can accelerate the graph processing, it can also lead to crashes when there is not enough memory for large amounts of messages or big datasets.

GPS [17]: GPS is also an open-source distributed graph-parallel processing system. GPS is a Pregel-like system, with three new improvements: First, in order to make global computation more easily expressed and more efficient GPS proposes an extended API with an additional function: *master.compute()*; second, GPS proposes a dynamic repartitioning scheme, which reassigns vertices to other computing nodes during the job execution based on messaging patterns; and last, GPS conducts an optimization, which distributes adjacency lists of high-degree vertices across all compute nodes, to further improve the system performance.

In summary, we have considered the diversity in platform type and computing model, and the representativeness and novelty in similar systems as selecting standard consequently selecting one advanced data parallel platform and three most representative graph parallel platforms to evaluate and analyze in this work.

3 Experimental Evaluation and Analysis

3.1 Experiment Environment

Our experiments were conducted on a Linux-based cluster with 65 nodes (528 processors). The cluster consists of one front-end node that runs the TORQUE resource manager and the Moab scheduler and 64 computing (worker) nodes. Each computing node has 16 GB RAM and 2 quad-core Intel Xeon 2.66GHz CPUs. All the nodes share /home partition through NFS. Up to 48 nodes were used in our experiments.

3.2 Experiment Design, Results, and Analysis

We obtained the results of the targeted metrics mainly through two methods: extracting information from job execution logs and using Linux ps command. In particular, the data processing time and volume of network I/O were obtained from the job execution logs. And the ingress time information on GPS and Giraph also came from their job execution logs. However, to measure the ingress time in Spark and PowerGraph, we modified the source code of these systems. We essentially measured the time spent between the initiation of a job and the beginning graph computing phase. Further, memory consumption and CPU load were derived by averaging the snapshot results measured using the ps [30] command in Linux. We ran each experiment at least three times.

3.2.1 Applicability

An interesting observation from our study is that not all the tested platforms were able to handle all the datasets in the experiments. Only PowerGraph could successfully process the largest graph, G1. All other platforms (Spark, Giraph, and GPS) crashed[2] while working on dataset G1. For PowerGraph, it also crashed when loading G1 under the asynchronous execution mode. The reason of the crashes is that the memory required by those systems for processing the graph exceeds the total amount of physical memory that the system can supply. In particular, GPS crashes when a large number of messages overwhelm the message queue kept by the message parser thread. For Giraph, the memory on one particular worker node is exhausted during the graph loading phase. For PowerGraph asynchronous engine execution mode, if the number of graph partitions is large, the messages exchanged among the partitions will explode and exhaust the memory of computing nodes. Spark

[2]We conducted experiments with 16, 32, 48 computing nodes and system crash occurs in all of those experiments.

crashes due to the overflow of the Java heap space during the shuffle phase. Even in each node we increased the size of heap space up to almost equal to the size of the memory capability of that node (16GB) the crash still persisted.

3.2.2 Data Processing Rate

We measured the time spent on processing the datasets. As mentioned in Section 2 we measured both the ingress time and the computing time for a graph computation job.

Ingress Time: The default graph loading strategy of Giraph and GPS is multi-loading, that is, each computing node loads its part in parallel. In contrast, the default graph loading strategy of PowerGraph is single-loading, in which the master reads the graph file from disk and distributes the graph to computing nodes. To be consistent with the graph loading strategy of Giraph and GPS, we configured PowerGraph to also use multi-loading strategy. Two data partitioning methods—random partitioning and oblivious partitioning—are available for PowerGraph and we tested both.

Figure 1 shows the ingress time for datasets G2 and G3 on each tested platform. There is no ingress time data available for G1 as our tested platforms crashed during data loading. We can clearly see in the figure that the ingress times of those graph-parallel processing platforms decrease when adding more computing nodes for medium-size graph G2 while they keep almost constant for small-size graph G3. For these graph-parallel processing platforms, the ingress time consists of the time for reading and partitioning the data and

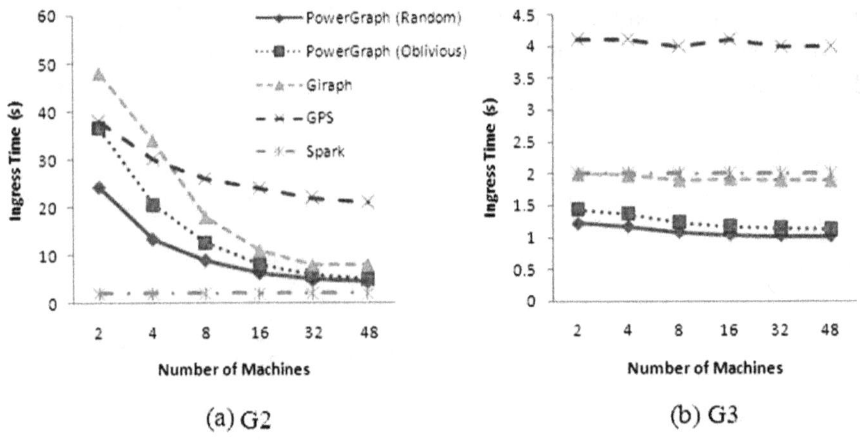

(a) G2 (b) G3

Figure 1 Ingress time for different dataset on each platform

the time for system setup. When the dataset is large and multi-loading strategy is employed, increase of nodes increases the degree of parallelism, effectively reduces the workload of data reading and partitioning on each node, and therefore reduces the overall ingress time. For small graphs, the graph loading time is negligible and ingress time is mainly determined by the system setup time, which stays approximately the same when more nodes are added.

Another interesting observation is that the ingress time of Spark is fairly small in all cases and its curve keeps flat for both G2 and G3. Unlike the three evaluated graph-parallel processing platforms that need to load the entire graph into memory before running any graph computation job, Spark does not need to wait for the whole graph being loaded into memory. Instead, Spark begins its data computing immediately after part of the data is available in memory. The ingress time of Spark actually records the time for loading partial graph instead of the entire graph. Thus the ingress time of Spark is small and remains constant even though more nodes are added.

Graph Computing Time:

Figures 2–4 show the computing time results of each platform for three selected graph algorithms with different dataset and number of computing nodes. Note that, the triangle counting algorithm implemented in PowerGraph does not support the asynchronous execution mode. Thus our results do not include those results of triangle counting under asynchronous mode in PowerGraph. Since all the platforms except PowerGraph were not able to handle G1 dataset, we only show the PowerGraph results on G1 in those figures. PowerGraph exhibits different execution times when it is configured with synchronous communication mode and asynchronous communication mode (the ingress times in both modes are the same). Due to the excessive

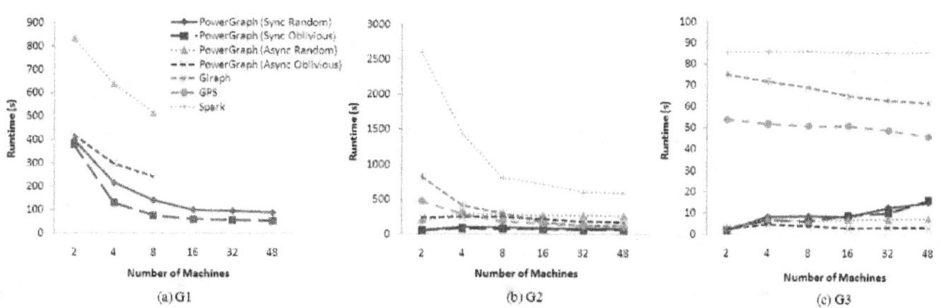

Figure 2 Computing time of PageRank

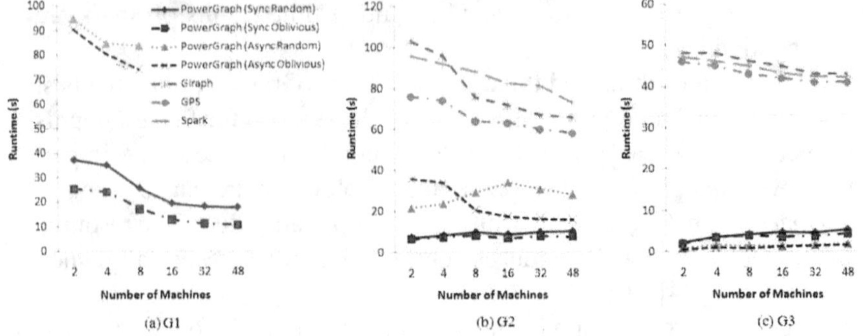

Figure 3 Computing time of single source shortest path (SSSP)

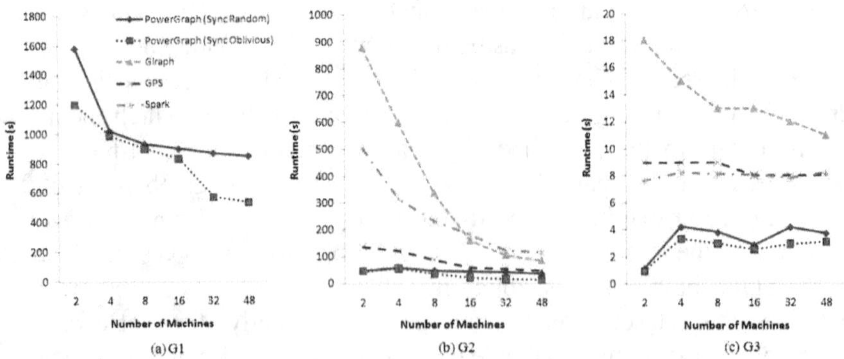

Figure 4 Graph computing time of triangle counting

memory usage incurred by asynchronous communications, we were only able to run PageRank and SSSP algorithms on PowerGraph in synchronous communication mode with G1 on 2, 4, and 8 computing nodes and show their results in Figures 2 and 3.

For PageRank, Spark always performs worse than other platforms in terms of computing time. The main reason is that Spark has a significant I/O between two continuous iterations and suffers a heavy shuffle phase between the map and the reduce phases. The implementations of single-source shortest path (SSSP) and triangle counting algorithms we used for Spark come from the graph analytic library supplied by Spark. The graph analytic library on Spark efficiently formulates graph computation within the Spark data-parallel framework, and it exploits the special graph data structures such as vertex cuts, structural indices and graph-parallel operators [24]. Thus, for SSSP and

triangle counting algorithms, the graph computing time results of Spark are more competitive.

For all three algorithms and three datasets PowerGraph exhibits the best performance in graph computing time. One of the reasons for PowerGraph's superior performance can be attributed to its highly optimized C++ implementation. All other platforms use Java as the development language. Another reason is that Giraph, GPS and Spark all use Hadoop MapReduce-like resource scheduling mechanism, which induces substantial overhead especially when the graph is small [24].

Because asynchronous mode incurs more communications than synchronous mode, PowerGraph (v2.2) cannot achieve runtime gain from asynchronous mode under the experiment environment of this work. For dataset G2 and G3 which are relatively small graphs, PowerGraph does not achieve better runtime when increasing the number of computing nodes. The runtime even increases when more machines are used. The reason is that PowerGraph splits the high-degree vertices across partitions, which incurs the overhead due to the joins and aggregations required for coordinating vertex properties across partitions [24]. When the size of graph is small, the synchronization overhead exceeds the benefit obtained from workload distribution. For the largest dataset G1 PowerGraph exhibits good scalability for all the algorithms. For other platforms, their scalability becomes more evident when the dataset is relatively large. Consequently, distributed data processing platforms are more suitable for processing large size of data, which is especially true for PowerGraph.

3.2.3 Resource Utilization

In order to understand the resource utilization of the targeted platforms, we collected network I/O, CPU load, and memory usage of the jobs executed on the selected platforms.

Network I/O: We measured the overall volume of network traffic (in unit of GB) generated by each platform when executing each of the three algorithms. Both G2 and G3 were tested. Results for PageRank, SSSP, and triangle counting are depicted in Figures 5, 6, and 7, respectively. As shown in those figures, in general, the more nodes are employed (i.e., the more partitions are generated), the more communications are incurred, which is reflected by rising curves in the figures. One notable exception to this pattern is the network traffic generated by Giraph, which remains fairly constant across different datasets and number of nodes for all three different algorithms. Relative lower communication overhead growth rate with more machines being used is one

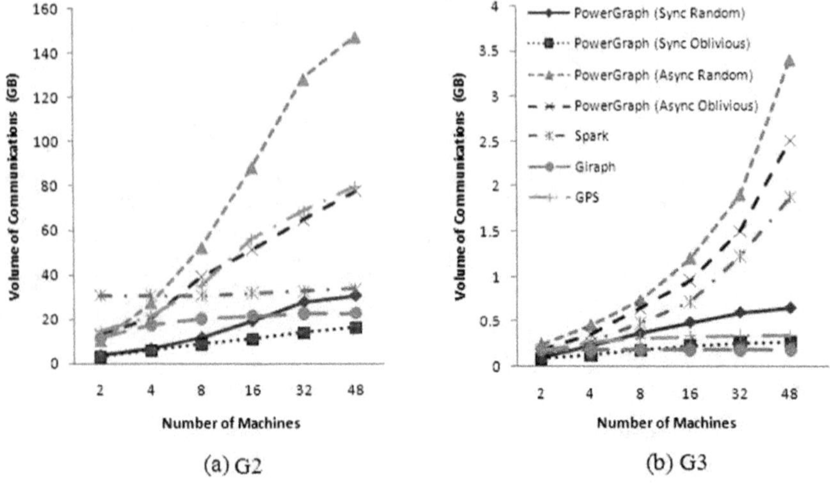

(a) G2 (b) G3

Figure 5 Network I/O of PageRank

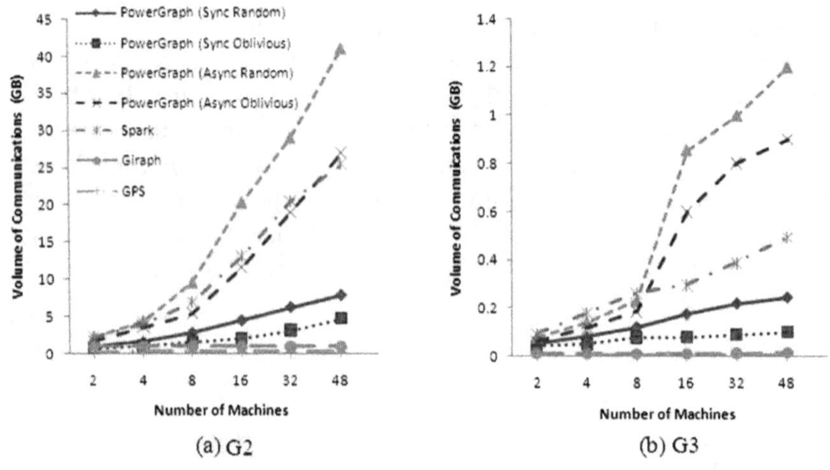

(a) G2 (b) G3

Figure 6 Network I/O of SSSP

of the reasons why Giraph can achieve a better scalability on graph computing time than other platforms.

Among the evaluated platforms the asynchronous PowerGraph suffers the largest communication overhead especially when the number of machines used becomes large. Because there is no synchronization between each round of computing, asynchronous PowerGraph needs to launch more

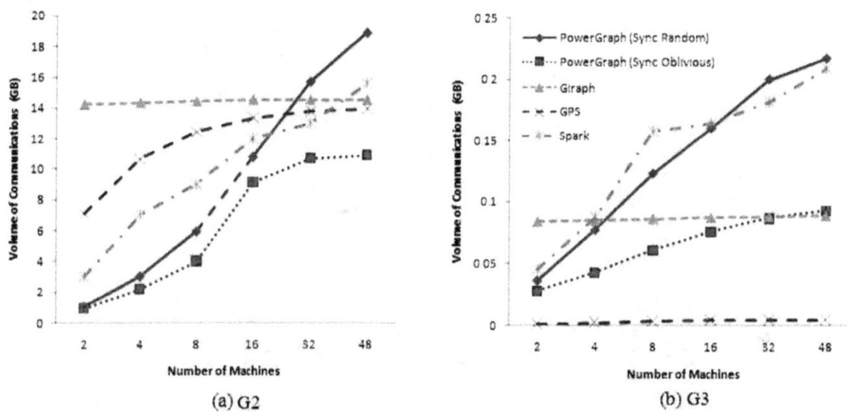

Figure 7 Network I/O of triangle counting

communications to synchronize the vertex properties across partitions, which may account for the fact that the graph processing rate of PowerGraph in the asynchronous mode is worse than that in the synchronous mode.

Memory Consumption: We measured the average memory consumption of each platform for running PageRank, SSSP, and triangle counting algorithms with datasets G2 and G3 and show their results in Figures 8–10. As shown in the figures, the more machines participate in the computing, the more memory each job will consume. Similar to the results in graph computing time, compared with other platforms, Spark consumes relatively more memory resource for PageRank algorithm. But its graph analytic library performs much

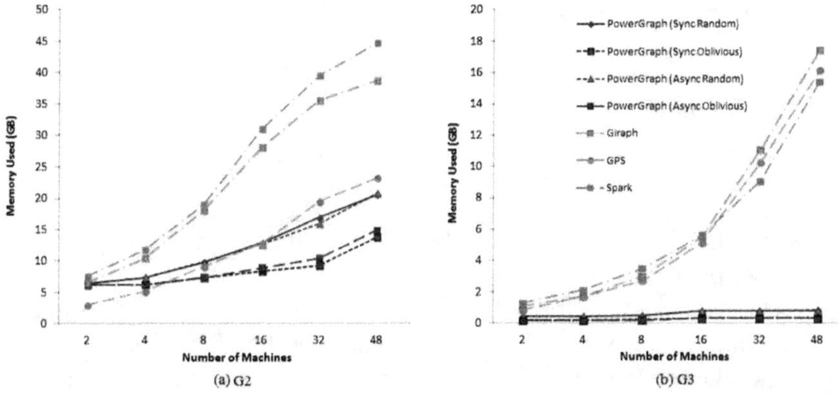

Figure 8 Memory used by PageRank

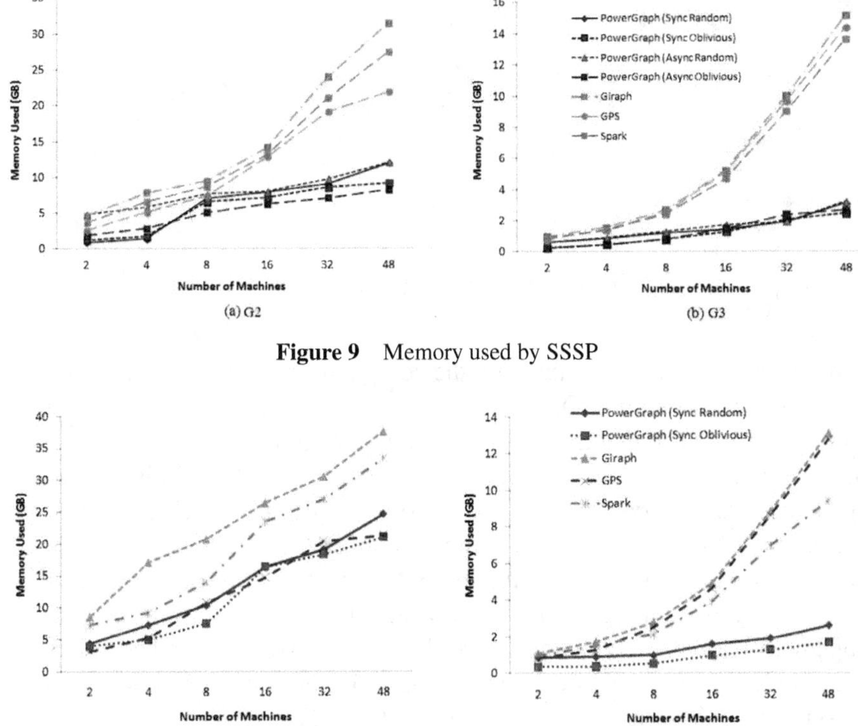

Figure 9 Memory used by SSSP

Figure 10 Memory used by triangle counting

better in memory usage. It is similar that the graph analytic library outperforms the raw Spark in the CPU resource utilization, which is presented in the following section. The reason is that the graph-parallel systems only conduct the operation related to the activated vertices in each iteration. This feature help to reduce the memory consumption for graph-parallel systems.

CPU Load: We measured the average CPU load of all computing nodes involved in the execution of each algorithm for each platform. In particular, we first sampled the CPU load of each computing node every second and calculated the average of the samples. Then, we averaged these CPU load values of all computing nodes. As the results of each platform exhibited similar trends for G1 and G2, only the results for G1 are presented.

Figure 11 depicts the variation of CPU load with the increase of computing nodes. It is clear that with the increase of the number of machines used, the CPU load of each machine goes down evidently. This is expected as more

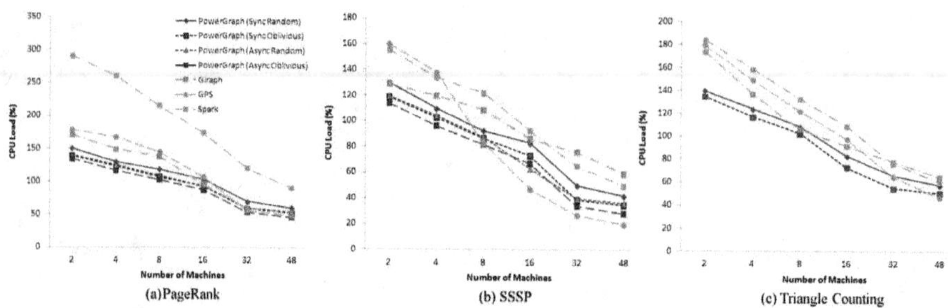

Figure 11 CPU Load for G2

machines process the graph in parallel, the computation workload distributed to each machine decreases. On the contrary, there is more CPU idle time induced by waiting for data transmission among machines.

Spark consumes more CPU time during job execution for all three algorithms (and for all datasets). One of the reasons for the high CPU load is that the ingress time for Spark is much shorter than those for other platforms. The operations during ingress time are mainly for loading the graph data from disk and configuring the running environment, which incur minor CPU utilization. Among graph-parallel processing systems, PowerGraph runs faster than others with lower CPU load. And its CPU load in asynchronous mode is lower than that in synchronous mode. The relatively low CPU load in asynchronous mode coincides with the relatively high communication overhead, which suggests that PowerGraph has more idle CPU in asynchronous mode due to more network communications occurred during graph processing.

4 Related Work

Many previous works [11, 15, 17, 24, 25, 26] have conducted performance evaluation of distributed graph-parallel computing platforms as well. Study [11] presented a detailed and real-world performance evaluation of six representative graph-processing systems (i.e., (Hadoop, YARN, Stratosphere, Giraph, GraphLab, and Neo4j) aiming to facilitate platform selection and tuning for processing tasks in Small and Medium Enterprise (SME) environment where computing resources are limited. In [15], the authors presented an evaluation of big data processing frameworks and provided a comparison between MapReduce and graph-parallel computing paradigms. However, its analysis is limited to the k-core decomposition problem. A number of other works [17, 24, 25, 26] have proposed new computing platforms and conducted evaluation of a few graph-parallel computing systems to prove the effectiveness of the new platforms. Thus, in terms of evaluation, these work are diverted from conducting a comprehensive evaluation and analysis of graph-parallel processing systems. In all these existing works, studies on the following two aspects are inadequate. First, in contrast to bulk synchronous computing model, asynchronous computing model has been recently proposed for graph processing. Existing works leave out thorough comparison and analyses of the performance exhibited by these two different computing models. Second, these existing works do not attach importance to explore and analyze the performance impact of the scale of datasets on distributed graph-parallel processing systems. Although some of their evaluations used different sized datasets, they failed to answer some interesting and important problems such as "Is it suitable to use a distributed computing system to process relatively small datasets?" and "What are the different performance characteristics when processing a large and a relative smaller dataset, respectively?". The answers to these questions can be valuable reference to users for platform selection. By conducting a comprehensive evaluation and analysis on the most concerned performance aspects of representative platforms, we seek to find the answers to these questions.

5 Conclusion

In this work, we performed a comprehensive evaluation of several popular graph-parallel computing platforms aiming to facilitate platform selection and tuning. For completeness, we also compared the performance of these graph-parallel systems with a popular data-parallel processing platform, Spark. We found that graph-parallel computing platforms outperform general data-parallel systems on both graph computing rate and resource utilization for processing graph-structured data. However, Spark uses less ingress time than graph-parallel systems when the sizes of the graphs are large. Further, all systems evaluated in this work exhibit better scalability on large dataset compared with small dataset. With an optimized C++ implementation and a sophisticated modular scheduling mechanism, PowerGraph exhibits a better graph processing rate than other graph-parallel computing platforms. Furthermore, PowerGraph has an intelligent graph partitioning strategy, which coordinates the vertices distribution among machines, improves both the execution efficiency and resource utilization. However, we found that the asynchronous computing mode proposed by PowerGraph does not bring improvement in performance in our experiments. Moreover, Giraph can achieve a better scalability on graph computing time than other platforms due to its relatively lower communication overhead growth rate when more machines are used. However, its graph computing rate is slower than PowerGraph and GPS.

6 Acknowledgment

This work was supported in part by the National Science Foundation under Grant CRI CNS-0855248, Grant EPS-0701890, Grant EPS-0918970, and Grant MRI CNS-0619069.

References

1. Forum, M.P., *MPI: A Message-Passing Interface Standard.* 1994, University of Tennessee.
2. Dagum, L. and R. Menon, *OpenMP: An Industry-Standard API for Shared-Memory Programming.* IEEE Comput. Sci. Eng., 1998. **5**(1): p. 46–55.
3. Dean, J. and S. Ghemawat, *MapReduce: simplified data processing on large clusters.* Commun. ACM, 2008. **51**(1): p. 107–113.
4. Low, Y., et al., *Distributed GraphLab: a framework for machine learning and data mining in the cloud.* Proc. VLDB Endow., 2012. **5**(8): p. 716–727.
5. Low, Y., et al., *Graphlab: A new framework for parallel machine learning.* arXiv preprint arXiv:1006.4990, 2010.
6. Gonzalez, J.E., et al., *PowerGraph: distributed graph-parallel computation on natural graphs*, in *Proceedings of the 10th USENIX conference on Operating Systems Design and Implementation.* 2012, USENIX Association: Hollywood, CA, USA. p. 17–30.
7. Katz, R.F., et al., *Numerical simulation of geodynamic processes with the Portable Extensible Toolkit for Scientific Computation.* Physics of the Earth and Planetary Interiors, 2007. **163**(1–4): p. 52–68.
8. Chen, W.-Y., et al., *Parallel spectral clustering in distributed systems.* Pattern Analysis and Machine Intelligence, IEEE Transactions on, 2011. **33**(3): p. 568–586.
9. Zaharia, M., et al., *Spark: cluster computing with working sets*, in *Proceedings of the 2nd USENIX conference on Hot topics in cloud computing*. 2010, USENIX Association: Boston, MA. p. 10–10.

10. Zaharia, M., et al., *Resilient distributed datasets: a fault-tolerant abstraction for in-memory cluster computing*, in *Proceedings of the 9th USENIX conference on Networked Systems Design and Implementation*. 2012, USENIX Association: San Jose, CA. p. 2–2.
11. Guo, Y., et al. *How well do graph-processing platforms perform? an empirical performance evaluation and analysis.*
12. Scott, J. and P.J. Carrington, *The SAGE handbook of social network analysis*. 2011: SAGE publications.
13. Newman, M., *Networks: An Introduction*. 2010: Oxford University Press, Inc. 720.
14. Malewicz, G., et al., *Pregel: a system for large-scale graph processing*, in *Proceedings of the 2010 ACM SIGMOD International Conference on Management of data*. 2010, ACM: Indianapolis, Indiana, USA. p. 135–146.
15. Elser, B. and A. Montresor. *An evaluation study of Big Data frameworks for graph processing*. In *Big Data, 2013 IEEE International Conference on*. 2013.
16. Guo, Y., et al., *Towards Benchmarking Graph-Processing Platforms.*
17. Salihoglu, S. and J. Widom. *Gps: A graph processing system.* in *Proceedings of the 25th International Conference on Scientific and Statistical Database Management*. 2013. ACM.
18. The Apache Software Foundation. *Apache Giraph*. 2014 cited 2014; Available from: http://giraph.apache.org/.
19. Page, L., et al., *The PageRank citation ranking: Bringing order to the web*. 1999.
20. Haveliwala, T., S. Kamvar, and G. Jeh, *An analytical comparison of approaches to personalizing PageRank*. 2003.
21. Langville, A.N. and C.D. Meyer, *Deeper inside pagerank*. Internet Mathematics, 2004. **1**(3): p. 335–380.
22. Watts, D.J. and S.H. Strogatz, *Collective dynamics of 'small-world'networks*. nature, 1998. **393**(6684): p. 440–442.
23. Minas Gjoka, Maciej Kurant, Carter T. Butts and Athina Markopoulou, Walking in Facebook: A Case Study of Unbiased Sampling of OSNs, Proceedings of IEEE INFOCOM '10, San Diego, CA, 2010.
24. Crankshaw, Daniel, Ankur Dave, Reynold S. Xin, Joseph E. Gonzalez, Michael J. Franklin, and Ion Stoica, The GraphX Graph Processing System.
25. D. Gregor and A. Lumsdaine, "The Parallel BGL: A Generic Library for Distributed Graph Computations," POOSC, 2005.

26. K. Kambatla, G. Kollias, and A. Grama, "Efficient Large-Scale Graph Analysis in MapReduce," in PMAA, 2012.
27. M. Zaharia, M. Chowdhury, T. Das, A. Dave, J. Ma, M. McCauley, M. Franklin, S. Shenker, and I. Stoica. Fast and interactive analytics over Hadoop data with Spark. USENIX; login, 37(4), 2012.
28. Yue Zhao, Kenji Yoshigoe, Mengjun Xie, Suijian Zhou, Remzi Seker, and Jiang Bian, LightGraph: Lighten Communication in Distributed Graph-Parallel Processing, in Proceedings of the 3rd IEEE International Congress on Big Data (BigData 2014), Anchorage, Alaska, USA, 2014.
29. POWER, R., AND LI, J. Piccolo: building fast, distributed programs with partitioned tables. In OSDI (2010).
30. http://en.m.wikipedia.org/wiki/Ps_(Unix)

Biographies

Yue Zhao received his B.S and M.S. degree in Computer Science and Technique in 2006 and 2009, respectively, from Jilin University. Currently he is a Ph.D candidate in Integrated Computing program at University of Arkansas at Little Rock. His research interest includes big-data analytic, distributed computing, High performance computing, and wireless network.

Kenji Yoshigoe is an Associate Professor in the Department of Computer Science and the Director of Computational Research Center (CRC) at UALR. He received his Ph.D. degree in Computer Science and Engineering from the University of South Florida. He is currently investigating the reliability, security, and scalability of various interconnected systems ranging from tightly coupled high performance computing systems to resource-constrained wireless sensor networks

Mengjun Xie is an Assistant Professor in the Department of Computer Science at the University of Arkansas at Little Rock. He received his Ph.D. degree in Computer Science from the College of William and Mary. His research interests include cyber security, information assurance, mobile computing, and big data analytics.

Suijian Zhou is a Postdoctoral Researcher in the Department of Computer Science at UALR. He received his Ph.D degree in Particle Physics from the Institute of High Energy Physics in China. He participated in the Atlas Grid Computing project at CERN and the IGE Grid Computing project in Sweden during the past few years. He is interested in the Big Data analysis and Cloud Computing techniques.

Remzi Seker is a Professor in the Department of Electrical, Computer, Software, and Systems Engineering at Embry-Riddle Aeronautical University at Daytona Beach, Florida. He received his Ph.D. degree in Computer Engineering from the University of Alabama at Birmingham. His research interests are safety and security critical systems and computer forensics. He is co-author of one of the first papers that was published on Mobile Phishing, and possible techniques for preventing it.

Jiang Bian received the M.S. degree in Computer Science in 2007 and his Ph.D. degree in Integrated Computing in 2010 both from University of Arkansas at Little Rock, Little Rock. He is currently an Assistant Professor of Biomedical Informatics at University of Arkansas for Medical Sciences, Little Rock. His research interest includes big-data analytic, network science, machine learning, and knowledge discovery and representation.

A Cached Registration Scheme for IP Multimedia Subsystem (IMS)

Lava Al-Doski[1] and Seshadri Mohan[2]

[1]*NIKSUN,inc 100 Nassau park Blvd. Princeton, NJ, 08540, USA*
[2]*Systems Engineering Department, EIT 519 University of Arkansas at Little Rock 2801 S University Avenue, Little Rock, AR 72204, USA*
lava.aldoski@gmail.com, sxmohan@ualr.edu

Received 17 August 2014; Accepted 11 September 2014
Publication 7 October 2014

Abstract

IP Multimedia Subsystem (IMS), an architectural framework for delivering multimedia services, was standardized by 3GPP/3GPP2. It is integrated with 4G and will most likely be with 5G and beyond, so as to enable wireless carriers to provide rich multimedia services, such as IPTV, chat, push to talk, and video conference. To use these services, users need to perform registration procedure with IMS, which will provide user information to the system. Registration is also performed when users move from one network to another. Due to the complexity of IMS and the increasing demands for these services, a key challenge IMS faces is to provide QoS to meet user requirements. One of the key performance factors is delay encountered in registration and service establishment. Also, users' mobility impacts the rate of registration and consequently the signaling generated that must be handled by the system. This work proposes and analyzes a cached registration scheme to reduce the delay associated with registration. The work also examines the effects of different mobility models on the application layer, in particular IMS. The work studies the impact of user movement patterns on the system and examines the impact of the proposed cached registration scheme on these patterns.

Keywords: IP Multimedia Subsystem, QoS, Mobility Models, IMS cached registration scheme.

Journal of Cyber Security, Vol. 3, 317–338.
doi: 10.13052/jcsm2245-1439.334

1 Introduction

IP Multimedia System has been standardized for the purpose of providing a services infrastructure for 3G, 4G and future wireless networks yet to evolve. IMS was created by 3GPP/3GPP2 to be part of the transition to the Next Generation Network (NGN). IMS offers a wide range of services including IPTV, Push-to-Talk, Web Services, and presence. IMS infrastructure consists of a collection of servers, each serving a dedicated task. These servers deploy a variety of protocols to set up and manage services. In order to avail of IMS services, users are required to register first with the system. The registration procedure also occurs whenever the users move from one network to another or from one IMS administrative domain to another within the same network. Due to the complexity of IMS signaling, delays are expected due to queuing and processing of signaling messages. Service providers face many challenges in guaranteeing QoS requested by the user. Several issues arise that impact the delivery of required QoS. One of these issues is the impact of users' mobility patterns. These patterns determine the handover rate that affects the system. As users move from one network to another, they may be handed over from one IMS services domain to another, which will initiate registration with the new IMS. This paper proposes and analyzes a scheme that can minimize the overall delay experienced by users due to the registration process and thereby facilitate the delivery of QoS required by the users.

Literature review identified several papers that analyze the delays encountered within the IMS infrastructure. In [1], the authors investigate end-to-end SIP delay in WiMax- UMTS, WiMax – WiMax, UMTS-UMTS network. The results demonstrate that the processing delay is a key contributor [2] the author implemented a different kind of analysis. While the IMS worked within CDMA2000 environment, the authors performed end-to-end session setup delay analysis by breaking it down along the various layers and calculating the delay in each layer taking into consideration the delay property of each layer. These analyses are crucial for real time services and interactive services. In [3], the authors have suggested caching the routing data in I-CSCF. The study indicated that by utilizing the I-CSCF cache, the average incoming call setup time can be reduced.

In the literature review analyses considering user mobility have assumed mainly two mobility models (random walk, fluid flow) as a generic model to study the effect of delay on a system. These models do not represent the whole picture of user mobility patterns and therefore fail in offering the best representation of the handover delay effects. Very few studies have been undertaken to explore the effects of different mobility models on the application layer. In this study, we propose a cached registration scheme for IMS users with mobility patterns associated with high handover rates and then analyze the impact of different mobility models on handover rate (registration rate) encountered by IMS. Ref. [17] provides an excellent analysis of mobility management from the perspective of system optimization.

The paper is organized as follows. Section II provides a brief overview of IMS architecture. Section III proposes a cached registration scheme, and considers several mobility models. Section IV describes the approach to modeling, simulation setup and an analysis. Section V discusses results and Section VI provides some concluding remarks.

2 IP Multimedia Subsystem Architecture

IMS consists of a set of core servers called Call Session Control Function servers, specifically, Proxy, Interrogating, and Serving Call Session Control Functions (P, I, S-CSCF). Each of these servers has its own functionality within the system. IMS also has its own data base to keep the users' authentication tokens and profile information. The focus of this paper is on the signaling load encountered by these servers and a scheme that serves to reduce the load. The P-CSCF serves as the first point of contact with the user and it is usually located in the visitor network. The I-CSCF mainly performs routing of the IMS messages. S-CSCF may be the considered the brain of IMS and the key decision maker. S-CSCF performs user authentication and registration, and oversees establishment, monitoring and tearing down of services. Both I-, and S-CSCF are located in the home network. IMS is connected to an assembly of servers denoted by application servers. These servers are responsible for providing the resource for IMS core servers to establish a service. Each server(s) is dedicated to a specific service. This collection can be part of IMS internal system or may belong to a 3rd party as shown in Figure 1 [4].

As shown in Figure 2, IMS users need to perform registration procedure with the system before establishing any service. This process establishes a connection with the user, loads user profile, performs authentication, allocates resources, and establishes security agreements [5]. The registration process proceeds as follows:

1. The user sends REGISTER to P-CSCF located in the visitor network. The proxy performs Domain Name System (DNS) query for the I-CSCF URI, and then forwards the REGISTER message.

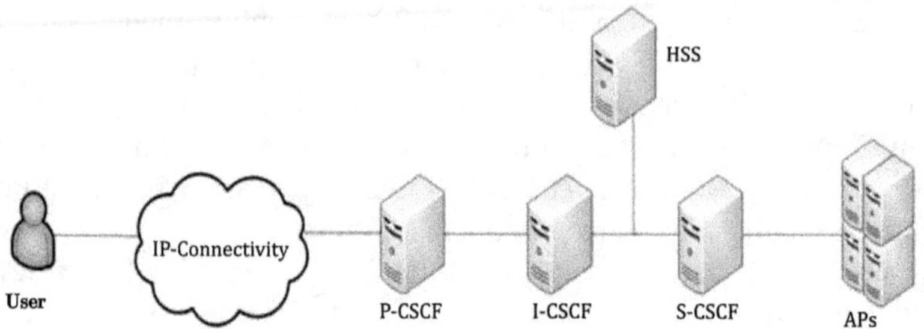

Figure 1 IMS architecture

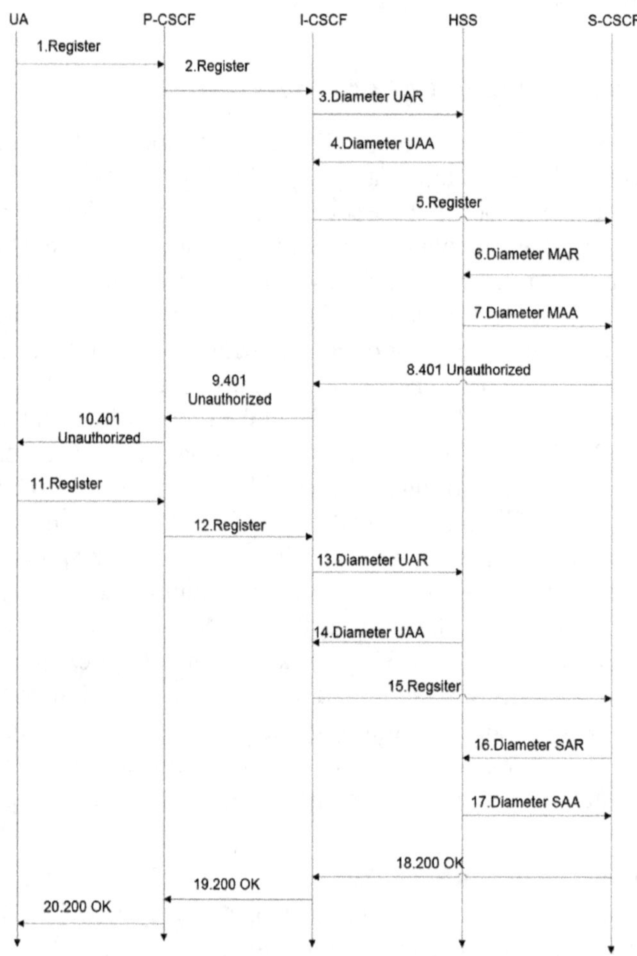

Figure 2 IMS registration procedure

The proxy inserts its address in the forwarded message to make sure that it receives back SIP response methods to its own address.

2. The I-CSCF serves as a router and a load balancer. The I-CSCF will examine the REGISTER message and extract the user public and private identities and network ID and then sends the information via Diameter protocol User-Authentication Request (UAR) to the HSS [6].
3. On receiving the message, the HSS processes the information and verifies the roaming agreement (in the case of receiving a visitor network ID).
4. The HSS authenticates the user and assigns a S-CSCF to the user. It then sends the response in a Diameter User-Authentication Answer (UAA).
5. The I-CSCF receives the UAA and extracts the S-CSCF address and then forwards SIP REGISTER to it.
6. The S-CSCF receives the REGISTER message and sends Multimedia Authorization Answer (MAA) to the HSS. The HSS stores the S-CSCF URI.
7. The HSS sends MAA back to the S-CSCF which contains authentication information of the user.

8, 9, 10- As part of the security system, IMS users are required to accept a challenge. Basically, IMS proposes a challenge for the user to respond, a way of protecting the system from flooding with fake users' request. So the user in this stage is not authorized.

11. The user sends a registration request again with the challenge response, which will be passed back through the same route.

12, 13- In the second attempt, the I-CSCF sends registration information to the HSS in the form of UAR.

14. The HSS sends back UAA which includes the address of the S-CSCF allocated to the user.
15. The I-CSCF sends the REGISTER to the S-CSCF with the user information.
16. The S-CSCF receives the user information and sends SAR to notify the HSS that the user is registered and authenticated.
17. The HSS receives SAR and sends the user profile, user public identity and the initial filter criteria to indicate the application server URI that would be involved with the user.

18, 19, 20- the S-CSCF sends 200 OK to indicate the registration procedure was successful.

3 Proposed Cached Registration Scheme

The volume of signaling associated with registration could reach substantial scales especially in crowded areas with hybrid networks triggering frequent handovers. These registration signaling can adversely impact the QoS.

When a service is interrupted or a handover is about to occur, the P-CSCF receives a graceful or ungraceful disconnect of service. This disconnect request is sent to the P-CSCF. The P-CSCF will inform the S-CSCF of the discontinuity of service through the I-CSCF. Subsequently, if the user's terminal performs a handover the user will send a registration request. The user might also send a re-establishment of a session if the session was interrupted during handover. Figure 3 depicts the flow diagram of the registration procedure where the user sends a registration request. In steps, the S-CSCF will send a request to the (MAR) HSS to download the user profile and save the

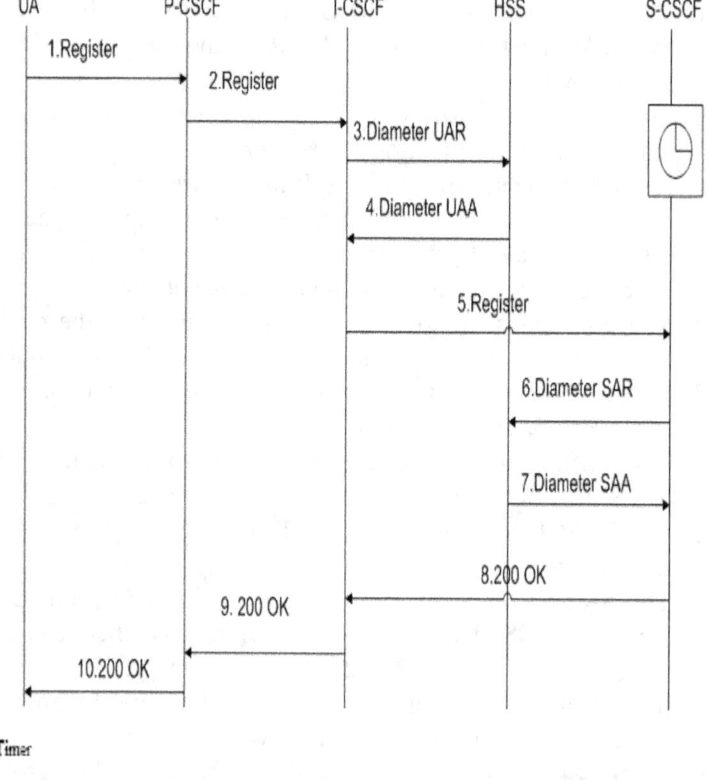

Figure 3 Cached registration scheme

S-CSCF address in the HSS [7]. After this the S-CSCF will send an unauthorized message to the user. The message includes a challenge to the user the challenge is part of IMS security features. In steps, the S-CSCF will send a SAR message to the HSS. If the authentication is successful, the S- CSCF will inform the HSS that the user is now registered and request to download the profile (see Appendix A).

The proposed caching scheme in this work suggests adding a timer to the S-CSCF, the timer is set to represent the time of the handover delay associated with the area of coverage of IMS. The S-CSCF will start the timer. While the timer is decreasing the S-CSCF will keep the user profile and authentication. When the timer expires the S-CSCF deletes the user profile and authentication. In this case, if a registration request is received a full registration scheme needs to be performed. Caching the user data during the life of the timer will cut down on the amount of registration signaling and therefore reduces the delay time. Considering the user authentication information is kept within the S-CSCF before the timer expires, the procedure suggests eliminating the security challenge proposed to the user. This procedure is suggested when the IMS coverage area has high traffic of registration requests arriving to the IMS system. In addition, it can be implemented in case of emergency, or spike change in user population in the area which affects the registration requests rate.

In order to capture and show the effects of different mobility models on the system, this work presents in what follows four case studies for a given area with a different scenario of mobility model in each case. The intention of this presentation is to bring to light the degree with which the IMS system is affected by the user's mobility pattern. Several realistic mobility models are utilized to represent users' mobility patterns in a coverage area. This work focuses on exploring mobility models that have not been considered before for delay effects on IMS. Users need to register every time a handover occurs in the system. Users with high mobility have higher registration rate. These requests increase the overall delay in the system and consequently degrade QoS. For this work the mobility models were generated using Matlab simulation tool. These scenarios incorporate the expected handover rate when users move from one networks to another in a hybrid network environment. The hybrid networks represent 4G networks (Wimax and LTE) loosely coupled with WLAN (see Appendix B).

Random Waypoint Model: The Random Waypoint Model is the most frequent model used in the simulations. The Random waypoint model is very similar to the Random Walk Model except that Random Waypoint Model has introduced the pause time parameter to the Random Walk Model. This is done in order to smooth out the node movement and prevent it from having sharp turns and changes in the velocity [8].

Random Walk Model: The Random walk mobility model is a user mobility model which is related to Brownian motion originally described mathematically by Einstein. It is used in different fields to resemble objects with random movements in a given area. The directions of the users and speed are selected randomly within a given range. For every new time slot, the new direction of the moving node is chosen from $(0, 2\pi)$ and the speed usually has a Gaussian distribution from $[0, V_m]$. The pause time for this model is zero. This model is memoryless since the direction and speed do not depend on future or past values [8], [9].

Manhattan Grid Model: Manhattan grid can represent the movement of users in a crowded city center. In the simulations it is assumed that buildings in the city center provide users with WiFi connections. Thus users need to perform handover when they step in or out of the WiFi into the 4G network. IMS in this case will be affected with handovers requests [10].

Fluid Flow Model: Users in the fluid flow model start at one point and then move in a uniformly distributed direction over $(0, 2\pi)$. They move with an average velocity of v and density of ρ and the area boundary length is L [11], [12]. The following equation gives R which represents the rate of registration.

$$R = \frac{\rho v L}{\pi}$$

4 Modeling, Simulation, and Analysis

IMS Model: IMS was modeled using the Arena® simulation tool [13]. A simulation model was built to identify and analyze the total delay experienced by users when requesting IMS services. The total delay includes the registration delay and the different services delays. To model the IMS system's proxy server, (usually located in the visitor network) the simulation model assumed 10 proxies spread over the entire coverage area. Since IMS services follow different flow diagrams, the services pose different loads on the main IMS servers. Although the registration process is not a part of the initially defined services, it was considered as such in the model since every user of the system needs to register with the IMS before requesting any service. The overall performance measure considered in this study involved the Total Delay (D_T) experienced by users when requesting IMS services. The total delay was defined as the summation of the Transmission Delay (D_t), Queuing Delay (D_q) and Processing Delay (D_p), as follows:

$$D_T = D_t + D_q + D_p$$

Where, the Transmission Delay was defined as the delay time which occurs when sending the packets from one end to another, the Processing Delay was defined as the time it takes for the servers to process a request or query, and the Queuing Delay was defined as the time the IMS message spends in the queue of the system. Only queuing and processing delay were considered in this work. Each of the IMS servers was modeled as a network of M/M/1 queues with feedback representing the message transfer between those servers. The processing rate had an exponential distribution with a mean of 5000. The applications layer and data base delay were not considered. All the results were obtained from 5 hour simulation periods with a warm up period of 20 minutes, each simulation was replicated 4 times. The results were within 95% confidence intervals. The Arena model reflects the amount of system delay experienced in the model compared to the different arrival rates.

Mobility Models Simulation: The work also focus on the effects of different mobility models on the IMS system. The effects of mobility patterns appeared in the form of handovers rate between networks. The user needs to perform handover when moving from one network to another. The simulation adopted an area covered by 4G network and few areas with WiFi. All the simulations have had the same assumed parameters in order to insure unified results. The following assumptions were made: the simulation area was taken to be 400 by 400 m; the number of users were 200; and velocity of the users varies among mobility models. The simulations were carried out for 1000 iterations. Each iteration was considered a time unit of 100. The simulation was carried out by reflecting the characteristics of the considered mobility models (Fluid Flow, Random Walk, Random Way Point, Manhattan Grid). These characteristics represent the nature of the movements, the nature of the environment, distribution of the networks and users behaviors etc. During the simulation when users move from one network to another, it will trigger a handover [14]. The handover counter increases by one. The average number of handover ($H_{average}$) is given by

$$H_{average} = \frac{1}{U} \sum_{k=0}^{U} H_i(k)$$

where U is the number of total iterations, and k, the kth experiment. The following section describes the mobility model algorithms implemented in the simulations together with the distribution of the handover for each model.

The proposed cached registration scheme focuses on reducing the queuing delay and hence the registration by caching the users' information for a given time. It trims down the signaling among IMS servers. The user profile and authentication information are cached in the S-CSCF. When the S-CSCF receives registration request and finds the user authentication and profile are available, it will send an update to the HSS for the registration and authentication. When the user information is cached the amount of signaling is reduced as shown in Figure 3. Both P-CSCF and S-CSCF have the same number of queries which is equal to 2. While the I-CSCF is equal to 3. Thus reduce the queuing delay associated with registration procedure [15] as shown

$$fW_q(t) = \mu(1-p)e^{-\mu(1-p)t}$$

$$\text{Let } \alpha = \mu(1-p)$$

$$fW_q(t) = \alpha e^{-\alpha t}$$

Where μ is the mean service rate, λ mean arrival rate and p is the utilization. Below it shows the total waiting time distribution function for each message in a flow diagram. The total waiting time distribution function is equivalent to the convolution of a waiting time distributions for each server query. The number of registration queries to P-CSCF and S-CSCF is equal to 4 and for I-CSCF the number of queries is 6 as seen in Figure 2. The queries between the user and P-CSCF are not included. In the registration flow some servers have more queries than others and the utilization function differs from one server to another.

$$p_1 = p_p = p_s = \frac{4\lambda}{0.005}$$

$$p_2 = p_i = \frac{6\lambda}{0.005}$$

$$\alpha_1 = \mu(1-p_1)$$

$$\alpha_2 = \mu(1 - p_2)$$

To solve the convolution between the pdfs, the Laplace transform of the equation as shown in [16] is:

$$L\{f_{WTotal}(s)\} = \left(\frac{\alpha_1^8}{(s+\alpha_1)^4}\right)\left(\frac{\alpha_2^6}{(s+\alpha_2)^3}\right)$$

Transforming it to the time domain shows the function of the total waiting time for every queue as follows:

$$L\{f_{WTotal}(s)\} = \left(\alpha_1^4 \frac{t^8}{7!} e^{-\alpha t}\right) * \left(\alpha_2^3 \frac{t^6}{5!} e^{-\alpha t}\right)$$

f_{WTotal} represents the total waiting time. The following equation also shows the total waiting time while using the proposed technique:

$$p_1 = p_p = p_s = \frac{2\lambda}{0.005}$$

$$p_2 = p_i = \frac{3\lambda}{0.005}$$

$$\alpha_1 = \mu(1 - p_1)$$

$$\alpha_2 = \mu(1 - p_2)$$

$$L\{f_{WTotal}(s)\} = \left(\frac{\alpha_1^4}{(s+\alpha_1)^4}\right)\left(\frac{\alpha_2^3}{(s+\alpha_2)^3}\right)$$

$$L\{f_{WTotal}(s)\} = \left(\alpha_1^4 \frac{t^4}{3!} e^{-\alpha t}\right) * \left(\alpha_2^3 \frac{t^3}{2!} e^{-\alpha t}\right)$$

Clearly, security of the system is an important factor and therefore must be suitably incorporated. The cached scheme may be introduced in special cases such as when the system experiences an unusual level of registration requests (example in case of emergency or sports events). The reduction in the delay increases when there is a substantial increase in the registration arrival rate that will not fit with the design of the IMS. To implement the scheme a timer needs to be added to the S-CSCF and a software modification to allow the users to register without the need to have security challenge during the time the timer has not expired.

Heretofore, the impact of different mobility models on the application layer has not received much attention. In this work, we investigate the degree with which different mobility patterns affect the application layer, in particular, the IMS signaling and server loads. The mobility patterns considered reflect different city environments, such as highways represented by fluid flow, city centers represented by the Manhattan grid, and pedestrians represented by random walk pattern. The impact of these mobility models is analyzed in the form of the rate of the expected number of handovers per second. The results were obtained from the simulation models created. Some mobility models have very little impact in comparison with other mobility models which have considerable impact on the system. In addition the cached registration scheme model is tested on these mobility models to see their effect in each environment. The cached scheme exhibits little to no effects on some mobility model and has significant impact on others as shown in the following section.

5 Results

From the created IMS model, the registration scheme was modeled to see the delay encountered due to the IMS servers while performing the registration procedure. Figure 4 shows the waiting time in each IMS server encountered by the signaling messages launched by the registration procedure for different arrival rates. The cached registration scheme is then implemented in the model. In comparison with IMS registration scheme without caching, the cached scheme provided a significant reduction in the delay experienced in the system. Figure 5 shows the delay in both schemes for different arrival rates.

While the cached registration scheme has shown an improvement in terms of delay, studying its impact on IMS with different mobility patterns provides the network designers an idea of when or where to implement it. The simulation results show the handovers distribution function for all scenarios. Each scenario exhibits a normal-like distribution function with different means and variances. Figures 6, 7, 8, and 9 show the distribution functions of Random Way Point, Random Walk, Manhattan Grid and Fluid Flow, respectively. Network designers will most likely benefit from these results by recognizing the capacity of the resources that need to be put in place to cater to the demands placed on the resources by each of these mobility models.

The results show the expected registration request delay in a city with different mobility patterns. The registration request arrival rate is not the only load factor on the servers, but other types of service requests are expected to

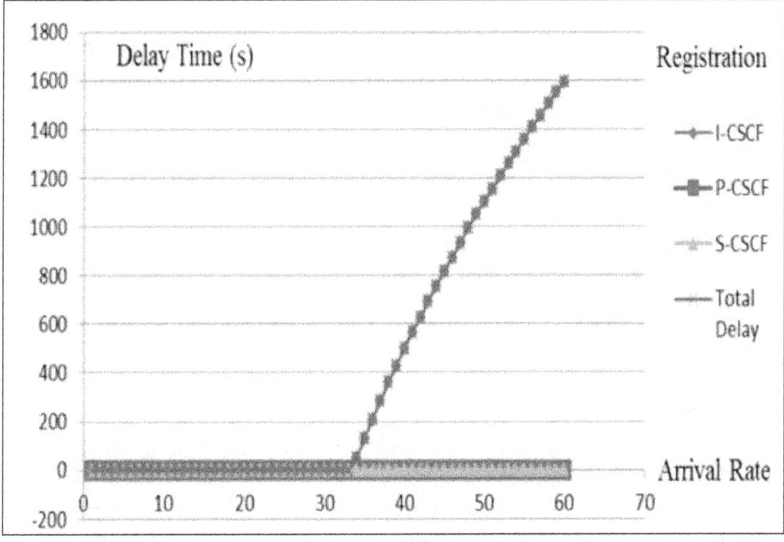

Figure 4 Registration delay (queuing and processing) in IMS servers

Table 1 Number of handovers/second for each mobility model

Mobility Model Type	Average Handovers/s
Random Walk	6
Fluid Flow	33
Random Way Point	3
Manhattan Grid	56

load the system. Service providers should study the area mobility different patterns representing a coverage area. From the distributions the average handovers per second may be calculated, which is given in Table 1 for different mobility models.

As already mentioned, users moving from one network to another network trigger handover, and, consequently, registration requests to the IMS. The expected handover of each model was used as an input for the IMS model to test the effectiveness of the cached registration scheme with mobility patterns of the user. Table 2 shows the delay associated with each mobility model on IMS system for the generic registration scheme and the proposed cached scheme. As can be seen in Table 2, random walk model produces an average handover of 6 per second while fluid flow produces 33 handovers per second, indicating that different mobility patters impact the handover rate differently.

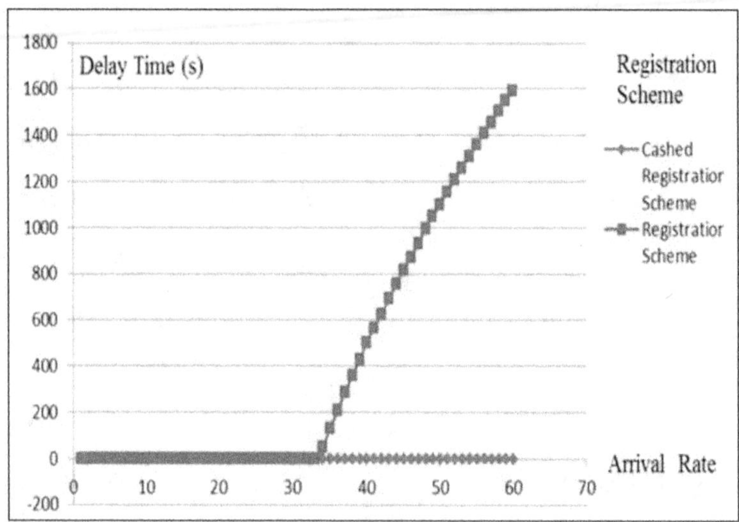

Figure 5 Registration scheme and cached registration scheme delay (queuing and processing)

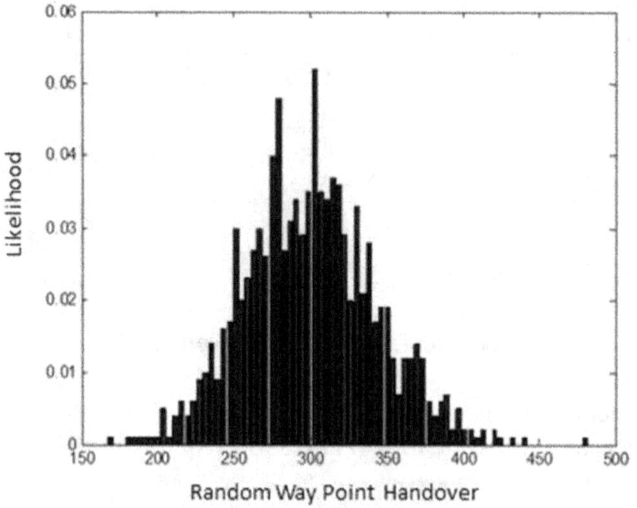

Figure 6 Random way point model handover distribution

Based on the results shown, service providers need to allocate enough resources for the registration rates in areas such as city centers and areas with a population of users with high mobility. The caching scheme can be very effective on areas with high handover rates and high mobility and may not

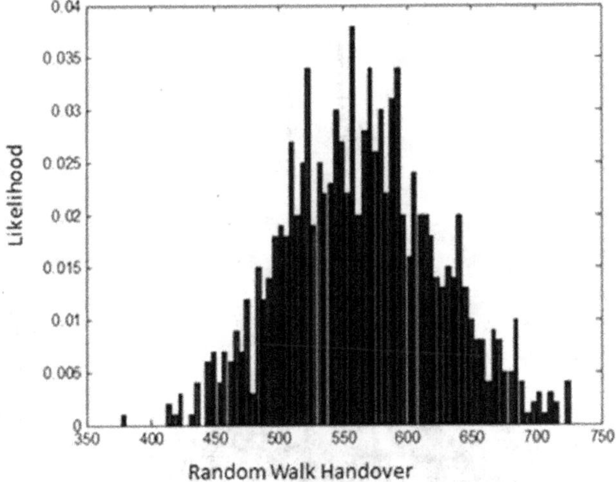

Figure 7 Random walk model handover distribution

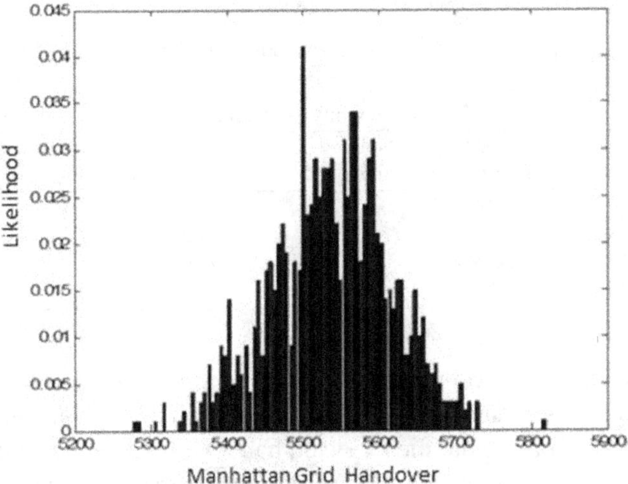

Figure 8 Manhattan grid model handover distribution

have substantial effects on users with mobility patterns similar to random walk and random way point. However, the impact of the cached scheme is more apparent in patterns similar to the fluid flow and Manhattan grid.

The mobility models considered would normally represent different aspects of a coverage area. Consequently, in most cases a single mobility model is not adequate to represent a service area. Thus, in a typical city with

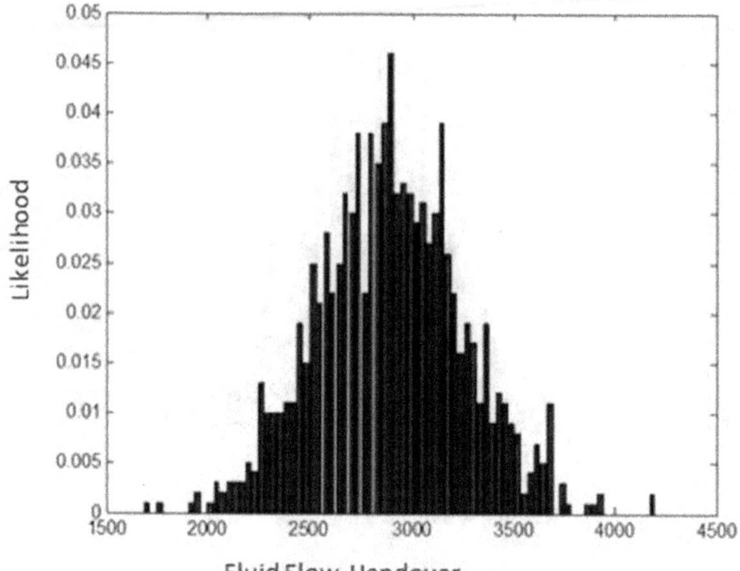

Figure 9 Fluid flow model handover distribution

Table 2 Handover delays for each mobility model for both registration scheme

Mobility Model Type	Delay with Default Registration Scheme	Delay with Cached Registration Scheme
Random Walk	0.0109889	0.000495
Fluid Flow	51.0	0.03
Random Way Point	0.004940	0.00023747
Manhattan Grid	1407.068	0.027

highways, city center, universities and airports different types of mobility patterns should be integrated for a good representation of the IMS serservice area of that city. Averaging out the expected handover rate per second for all four mobility models and applying this average value to the IMS model. Table 3 shows the Mobility Models handover average with the delay associated with the default registration scheme and the cached registration scheme.

Table 3 Mobility models average handover and delay for each registration scheme

	Average Handover/Second	Default Registration Delay in Seconds	Cached Registration in Seconds
Average Mobility Models Handovers	25	0.02	0.0046724

6 Conclusion

IP Multimedia Subsystem has shown promise to be the service infrastructure for next generation wireless networks. It caters to various services to the users, provides the environment for services development and 3rd party integration. IMS is a complex system, and service providers face challenges in providing QoS users may require. This work has focused on the analysis and reduction of delay associated with handover for different mobility models. The work shows the diverse effects accompanied with different mobility models. It was concluded that the impact on the system was significant in the case of the Manhattan grid and fluid flow models as compared to the random walk and random way point patterns. Thus service providers need to consider the mobility factor in areas such as city centers, where higher numbers of handovers are expected in a given time. For delay reduction, the cached registration scheme was simulated to demonstrate that in cases where high handover rate is expected, the scheme reduces the waiting time significantly. However, with random walk and random way point, the scheme bears little to no effects in comparison to the default scheme. Considering that in most cases the service area might have multiple patterns given, an adaptive scheme might work best that may apply either the default scheme or the cached scheme depending on the mobility pattern encountered in the area. A formal development of such an adaptive scheme and its analysis remains a topic for further research.

7 Appendix

Appendix A

Pseude Code: Cached Registration Scheme.

1- *User discontinues service with IMS.*
2- *S-CSCF keeps user information and authentication, S-CSCF timer start.*
   Timer =Timer-1
3- *Repeat until receiving registration request from same user OR Timer = 0.*
   If Timer =0.
   Delete user information and authentication. Perform complete registration.
   Else if Timer is not 0 perform cached registration.

Appendix B

Pseude Code: Mobility Models.

1- *Create 4G and wifi Network areas.*

2- *Assign initial locations for the users, the locations are chosen randomly.*

3- *Based on the initial locations retrieve users network type.*

4- *For each user*

If the user mobility pattern is Random Way Point Mobility

Do pause random selection between the values {0,2}

f pause is not zero then

User_pause = User_pause - 1

Else pause is zero

For user's velocity Do random selection between {-10,10}.

New user location (x,y)= Random selection{-10,10} + current user location (x,y)

Retrieve network type from users new location(x,y)

If change in the network type is detected from the previous location then

Handover= handover+1

If the user mobility pattern is Random Walk Mobility

For user's velocity Do random selection between {-10,10}

New user location (x,y)= Random selection{-10,10} + current user location (x,y)

Retrieve network type from users new location(x,y)

If change in the network type is detected from the previous location then

Handover= handover+1

If the user mobility pattern is Manhattan Grid Mobility

Do random selection of the user next step between {-3, 3}.

New user location (x,y)= Random selection{-3,3} + current user location (x,y)

Retrieve user's network type.

If change in the network type is detected from the previous location

Handover= handover+1

If the user mobility pattern is Fluid Flow Mobility

Do random selection of the user next step between.

New user location (x,y) = Random selection{-3,3} + current user location (x,y)

If New user location (x,y) > location (x,y) from the center of area.
Then assign the new locations to the user
Else repeat 5 until New user location (x,y) > location (x,y) from the center of the area.
If change in the network type is detected from the previous location
Handover= handover+1
5-Repeat 4 for the next user

References

[1] M. Ulvan, and R. Bestak "Delay Performance of Session Establishment Signaling in IP Multimedia Subsystem" 16th *International Conference on Systems, Signals and Image Processing*, pp 1–5, 2009.

[2] M. Melnyk, A. Jukan, and C. Polychronopoulos "A Cross-Layer Analysis of Session Setup Delay in IP Multimedia Subsystem (IMS) With EV-DO Wireless Transmissiona" *IEEE transactions on Multimedia*, vol 9, pp 869–881, 2007.

[3] Y.Lin, M. Tsai "Caching in I-CSCF of UMTS IP Multimedia Subsystem"*IEEE Transactions on Wireless Communications,*Vol 5, 2006.

[4] L. Al-Doski, A. Abbosh, R. Babiceanu, and S.Mohan. "Optimizing Resources Utilization in IP Multimedia Subsystems" *IEEE International Systems Conference (IEEESysCon)*. 2012.

[5] G. Camarillo and M. Garcia-Martin "The 3G IP multimedia subsystem" *John Wiley and Sons*, 2nd ed., February 2006.

[6] E. Fogelstroem, A. Jonsson, and C. Perkins "Mobile IPv4 Regional Registration. Internet draft, draft-ietf-mip4-reg-tunnel-04" *IETF*, work in progress, 2007.

[7] 3rd Generation Partnership Project (3GPP). *IP Multimedia Subsystem (IMS), TS23.228*, Rel10.

[8] J. Harri, F. Filali, and C. Bonnet "A Survey of Mobility Models for Ad Hoc Network Research" *IEEE Communication Surveys and tutorials,* volume 11, pp 19–41, 2009.

[9] W. Wang, and I. Akyildiz "Intersystem Location Update and Paging Schemes for Multitier wireless Networks" *Proceedings of the* 6th *annual international conference on Mobile computing and networking*, pp 99–109, 2000.

[10] F. Bai and A. Helmy "A Survey of Mobility Model" Book Chapter, *Kluwener Academic Publishers*, USA.

[11] L. Saleem, R Ghimire, and S. Mohan "An Analysis of IP Multimedia Subsystems for a Hybrid Network" *In IEEE* 4th *International Conference on Internet Multimedia Services Architecture and Applications (IMSAA)*, pp 1–6, 2010.

[12] S. Mohan, and R. Jain "Two User Location Strategies for Personal Communications Services" *In IEEE Personal Communications*, vol1, pp 42–50, 1994.

[13] D. Kelton, R. Sadowski, and D. Sturrock "Simulation with ARENA" *McGraw Hill*, ISBN 0-073-37628-0, 2003.

[14] L. Saleem and S. Mohan "A Complexity Analysis of IP Multimedia Subsystems (IMS)" *In First International Symposium on Advanced Networks and Telecommunications Systems (ANTS)*, December 2007.

[15] A. Papoulis, et al "Probability Random Variables and Stochastic Processes" *McGraw Hill*, ISBN 0-073-66011-6, 2002.

[16] A. Amooee and A. Falahati "On Total Call Setup Waiting Time Probability Distribution in IMS Signaling Network" 3rd *International Conference on Next Generation Mobile Application and Services*, pp 33–38, 2009.

[17] A. Dutta, "System Optimization for Mobility Management," PhD Dissertation, Columbia University, 2010.

Biographies

Lava Al-Doski Lava obtained both her PhD in Systems Engineering (Telecommunications and Networking Track) in May 2012, and Master of Science in Applied Science in December 2011, from the University of Arkansas at Little Rock. She received her Bachelor of Computer Engineering Technology in July 2005 from the Technical College of Mosul, Mosul, Iraq. She was employed as a student intern at AT&T, Atlanta, GA, for two semesters during 2010/2011, where she worked on IMS and VoIP. She is currently employed at Niksun as Mobility System Engineer. She is a member of IEEE since 2008.

Seshadri Mohan is currently a professor in Systems Engineering Department at University of Arkansas at Little Rock, where, from August 2004 to June 2013, he served as the Chair of the Department of Systems Engineering. Prior to the current position, he served as the Chief Technology Officer with Telsima, Santa Clara, California; as Chief Technology Officer with Comverse, Wakefield, Massachusetts; as a Senior Research Scientist, with Telcordia, Morristown, NJ and as a member of the technical staff with Bell Laboratories, Holmdel, NJ. Besides his industry positions, he also served as an associate professor at Clarkson University and as an assistant professor at Wayne State University. Dr. Mohan has authored/coauthored over 100 publications in the form of books, patents, and papers in refereed journals and conference proceedings. He has co-authored the textbook Source and Channel Coding: An Algorithmic Approach. He has contributed to several books, including Mobile Communications Handbook and The Communications Handbook (both CRC Press). He holds fourteen US and international patents in the area of wireless location management and authentication strategies as well as in the area of enhanced services for wireless. He is the recipient of the SAIC Publication Prize for Information and Communications Technology. He is currently serving as the Editor-in-Chief of Advances in Network and Communications. He has served or is serving on the Editorial Boards of IEEE Personal and Nomadic Communications (now IEEE Wireless Communications), IEEE Communications Surveys and Tutorials, and IEEE Communications Magazine and has chaired sessions in many international conferences and workshops. He has also served as a Guest Editor for several Special issues of *IEEE Network, IEEE Communications Magazine, and ACM MONET.* He served as a guest editor of March 2012 IEEE Communications Magazine feature topic titled "Convergence of Applications Services in Next Generation Networks" as well as the June 2012 feature topic titled "Social Networks Meet Wireless Networks." In April 2011, he was awarded *2010 IEEE Region 5 Outstanding Engineering Educator Award.* His paper titled "A Multi-Path Routing Scheme for GMPLS-Controlled WDM Networks," presented at the 4th *IEEE Advanced Networks and Telecommunications Systems Conference,* received the best paper award. Dr. Mohan holds a Ph.D. degree in electrical and computer engineering from McMaster University, Canada, the Master's degree in electrical engineering from the Indian Institute of Technology, Kanpur, India, and the Bachelor's (Honors) degree in Electronics and Telecommunications from the University of Madras, India.

Personal Denial of Service Attacks (PDOS) and Online Misbehavior: The Need for Cyber Ethics and Information Security Education on University Campuses

Ashley Podhradsky[1], Larry J. LeBlanc[2] and Michael R. Bartolacci[3]

[1]*Dakota State University, Madison, SD, USA*
[2]*Owen Graduate School of Management, Vanderbilt University, Nashville, TN, USA*
[3]*Pennsylvania State University, Berks, Reading, PA, USA*
ashley.podhradsky@dsu.edu,
larry.leblanc@owen.vanderbilt.edu,
mrb24@psu.edu

Received 14 August 2014; Accepted 11 September 2014
Publication 7 October 2014

Abstract

The authors examine the need to provide basic information security and cyber ethics training for all university students, not just those pursuing an information security-related degree. The authors also discuss the need to include ethical hacking, as part of an emphasis on cyber ethics, into information security degree programs. Both of these topics are discussed within the context of a new category of cyber crime, a Personal Denial of Service Attack (PDOS) that the authors have identified, along with other types of cyber crime, that are endemic to university campuses.

Keywords: Information security training, Personal Denial of Service Attack, Cyber Ethics.

Journal of Cyber Security, Vol. 3, 339–356.
doi: 10.13052/jcsm2245-1439.335

1 Introduction

The Internet has undeniably become a necessity for many people across the globe. From conducting online banking and paying bills, searching Google, and asking WebMD for medical advice, the Internet is relied upon for a myriad of functions in everyday life. Businesses and governments alike have the same dependency on Internet-based communication, cloud storage, online transaction processing capabilities, and other Internet-enabled tools to keep their entities up and running. With the vast volume of online business transactions today, the importance of securing our online world is more important than ever. Since cyber-attacks have become a fact of life as the Internet grows. Fred Cohen, who is best known for his early work on computer viruses, and Robert Tappan Morris, who created one of the first computer worms, have demonstrated the limitations and vulnerabilities of our networks [1]. Michael Glenn Mullen, retired U.S. Navy Admiral, stated that "the single biggest existential threat that's out there is cyber [2]." Given the importance of deterring and resolving cyber security attacks of varying degrees in the U.S., information security has become a top priority for corporations, government entities, and the general populace. In response to the need for information security education, the U.S. government has created the Comprehensive National Cybersecurity Initiative (CNCI) to expand the nation's cyber security educational capabilities [3]. Initiative #8 of the CNCI identifies that "Billions of dollars are being spent on new technologies to secure the U.S. Government in cyberspace; it is the people with the right knowledge, skills, and abilities to implement those technologies who will determine success." The initiative further adds that there are not enough cybersecurity experts and "we must develop a technologically-skilled and cyber-savvy workforce and an effective pipeline of future employees".

As a result, many universities are beginning to implement focused information security programs and are adapting existing curricula to focus on this important area based on standards developed by the National Security Administration [4]. The problem with the creation of just these limited scope programs, from the authors' point of view, is twofold:

- they limit the scope of information security education to a small set of students
- they do not include a sufficient coverage of cyber ethics, instead focusing on more traditional defensively-oriented information security skills

Cyber ethics must also play an equally important role in in cyber security education for all students at a university and would play a crucial role in their personal information security on campus. This work examines the need for information security and cyber ethics training for all university students and does so in the context of a newly categorized type of cybercrime, the Personal Denial of Service attack (PDOS) [5] and similar forms of online misbehavior. The need for basic information security and cyber ethics training for all students arises from the environment of a highly networked university campus and the close proximity students have with one another in it. The authors posit that this category of Internet crime, the PDOS, and many similar types of online misbehavior are ideally suited for universities' relaxed and social atmosphere of interaction and unlimited Internet connectivity. A PDOS is a cyber-crime in which an individual deliberately prevents the access of another individual or small group to online services. Such an attack can be undertaken using easily obtained information about a target's online services and Internet habits. We developed a survey was and administered it to students at four universities and a fifth non-university control group to help assess student attitudes towards online account security regarding a specific type of cyber crime, an online account breach. This survey provides some evidence that account breaches do occur regularly on university campuses to unsuspecting students and that existing laws do not deter such activity.

Although the initial impetus for this work was to examine the need to include basic information security and cyber ethics training for all students due to their vulnerabilities to attacks such as a PDOS, the authors also realized that information security curricula in the U.S. fell short of our expectations for its cyber ethics content. The scope of our work expanded to also include an examination of cyber ethics in the context of teaching ethical hacking. We argue that cyber ethics must not only be taught in the context of personal information security for the student body as a whole to limit the impact of online misbehavior, but must also exist in specific information security programs and their courses.

2 Student Online Misbehavior

The reliance on the Internet across the world has created a tech-savvy generation of young people who spend a good deal of their time online. "Millennials", as they are called, have grown up with online services such as Facebook, Google, Massively Multiplayer Online Games (MMOG's), online chatting, and social networks as an integral part of their lives. Given such a full time connection to online resources through a wide variety of Internet-connected devices, many users have been tempted to perform activities outside the boundaries of acceptable online behavior. The darker side of online activity includes more serious offenses such as cyber bullying, online fraud, and hacking, but also includes other types of online misbehavior that are sometimes overlooked or even tacitly accepted by this demographic group. The potential reasons for initiating these activities are myriad, but Routine Activities Theory [6] has been put forth to help explain the origins of crimes such as these. A component of the Routine Activities Theory is the assumption that anyone may commit a crime if given the opportunity or circumstances to do so. An additional assumption that relates to this is that victims of such crimes consciously placed themselves in situations where such crimes may occur. These notions, although controversial to some sociologists and criminologists, set the stage for the discussion of Personal Denial of Service Attacks (PDOS) and other lesser known forms of online misbehavior and their ramifications for students in a university setting.

As previously mentioned, a PDOS [5] is an attack on a person or small group where access to online services is denied through a clever manipulation of existing security procedures implemented by the online service providers. With the dependence on "the cloud" for accessing remotely hosted applications, synchronizing applications between devices, storage, and a myriad of other purposes, continuous access to online services accounts is not a mere luxury, but a necessity for many individuals to live their life each day. A PDOS intentionally seeks to invoke online account security procedures in order to deny a rightful account owner access to the account. This type of cybercrime, while falling short of an actual account breach or what is traditionally defined as cyber harassment, still represents a potentially damaging form of online misbehavior that is relatively easy to perpetrate and difficult to succinctly track. Universities are full of Internet-savvy young people, many with a "gaming" mentality, advanced online technical knowledge, and underdeveloped ethics, who are prime candidates to commit a PDOS attack or similar type of cyber crime. Other types of cyber crime such as theft of intellectual property, the illegal downloading and distribution of copyrighted material, and cyber harassment occur with great regularity on university campuses. Given this environment, and driven by the notions of the Routine Activities Theory, it is more prudent than ever to include basic information security and cyber ethics training as part of the curriculum for all university students.

According to the Privacy Rights Clearinghouse [7], cyber-stalking includes the list of actions detailed below. It should be noted that a PDOS is not directly described by any of the actions in this list. A PDOS attack

does not include sending threatening emails, hacking into an account (in fact in a PDOS, the attacker is intentional not breaching the account in order to activate security mechanisms by the online service provider), creating false accounts, posting messages to online boards, or spamming the victim. Therefore, a PDOS is one form of online misbehavior that university students might engage in with limited fear of legal oversight and policing.
Cyber-stalking actions:

- Sending manipulative, threatening, lewd or harassing emails from an assortment of email accounts.
- Hacking into a victim's online accounts (such as banking or email) and changing the victim's settings and passwords.
- Creating false online accounts on social networking and dating sites, impersonating the victim or attempting to establish contact with the victim by using a false persona.
- Posting messages to online bulletin boards and discussion groups with the victim's personal information, such as home address, phone number or Social Security number. Posts may also be lewd or controversial – and result in the victim receiving numerous emails, calls or visits from people who read the post online.
- Signing up for numerous online mailing lists and services using a victim's name and email address [7]

A PDOS, due to the fact that it is not attempting compromise the integrity of information contained in an online account, would generally fall under the category of cyber ethics for basic information security education on university campuses. It is important to insure that students understand that even an attempt to interfere with the online account of another represents a cyber ethics breach and maybe in fact be a crime in certain jurisdictions. While an information security curriculum teaches students how to deal with the ever-increasing number of ways that systems, networks, data, and organizations can be attacked electronically and how to manage the information security function in an organization, it is very limited in the scope of students that it reaches. We are stressing the need for all students to be given training in best practices for conducting the online portion of their lives which includes basic personal information security and cyber ethics content. Additionally, most information security programs at universities focus on the technical and managerial aspects of information security and do not stress the more ubiquitous societal attitudes towards online conduct and cyber ethics.

3 Need for Cyber Ethics in Information Security Curricula

In addition to the need for basic information security and cyber ethics training for all university students, a concerted effort is needed to incorporate more cyber ethics into information security curricula. While there is an increasing need for cyber security experts, there are limited qualified graduates [8]. There has been a recent increase in such programs in U.S. universities to address this need, but unfortunately the programs are strictly designed around the needs of corporations and government rather than person information security and cyber ethics.

The following is a brief introduction to typical information security curricula in the U.S. which shows that the emphasis is on technical and managerial skills training and not on the ethics of everyday Internet use and interaction with other users of online services. The National Security Administration (NSA) has two core academic accreditation programs: the Cyber Operations Program and the new Cyber Defense Program [8, 9]. These programs replace the existing Center of Excellence in Information Assurance Education (CAE IAE) and Center of Excellence in Information Assurance Research (CAE IAR) [9]. The Cyber Operations designation is seen as a symbol of excellence in offensive-based university curriculum. The program specifies both mandatory and optional knowledge units. Within the accreditation criteria, the university must include certain mandatory knowledge units, and 60% of the optional program content. Of all the required and optional knowledge units for this highly coveted program accreditation, ethics is not specifically listed in the academic content requirements, thus giving it a secondary role in the curriculum at best. Of the mandatory content, accredited programs must demonstrate coverage of low level programming, software reverse engineering, operating systems theory, networking, cellular and mobile communications, discrete math, an overview of cyber defense, information security fundamental principles, vulnerabilities, and legal topics. The optional program content includes programmable logic languages, FPGA design, wireless security, virtualization, large scale distributed systems, risk management of information systems, computer architecture, microcontroller design, software security analysis, secure software development, embedded systems, forensics, systems programming, applied cryptography, SCADA systems, HCI/Usable Security, Offensive Cyber Operations, and hardware reverse engineering.

The content in information security classes taught is always an open topic for discussion. Universities seek to isolate their exposure from the potential harm such content can wreak on their networks and any liability for cyber crime committed by students in such classes. Any wrong application of the skills learned in such classes may result in very serious consequences (such as expulsion from school, criminal cases, even the cancelation of academic programs, etc. [10]. Cook et al. [10] conducted a series of interviews in eight information security teaching schools at the graduate and under graduate level and discussed real scenarios in which students have misused their cyber security expertise. The research stresses that the students should feel ethical responsibility about application of ethical hacking techniques outside the classroom, but unfortunately such content is overlooked in favor of technical and managerial skill sets. From their interviews, Cook, et al have suggested eleven guidelines from 'hard learned lessons' which include: '(a) providing appropriate context and ethical tone, (b) explain the downsides of inappropriate behavior, (c) policies should be unambiguous, enforced, and legally defensible, (d) encourage students to pause and reflect before acting, (e) avoid stupid mistakes, (f) provide an ethics lesson early in the program, (g) tell the positive story of the program before something bad happens, (h) make students part of the process, (i) provide safe environments for exploration and experimentation, (j) provide leadership, mentorship and role models, and (k) don't crush the enthusiasm of the students". [10]

One methodology for incorporating cyber ethics into an information security program is the introduction of students to ethical hacking techniques. With the growth of the Internet from computers to phones to tablets there is an increase in the types of security threats around the world, challenging the academic world to prepare qualified information security and information assurance professionals (graduates) who also incorporate ethics into their skill sets [11–13]. Such techniques build a sense of responsibility for the security of all online users as a whole and teach students to find vulnerabilities in a responsible manner, including the reporting of those vulnerabilities to the appropriate parties. Teaching students how to attack systems and networks for the purpose of reporting vulnerabilities creates well rounded and ethical hackers who know more than just defensive techniques for information security [14–16]. Trabelsi and Ibrahim [17], researchers from the United Arab Emirates, have created a case study on "implementing comprehensive ethical hacking hands-on lab exercises" which include three Denial of Service (DOS) attacks: the TCP SYN flood attack, the Land attack, and the Teardrop attack. Trabelsi and Ibrahim have also suggested eight steps for educators and academic organizations (that deal offer information security courses) to minimize their liability related to student misbehavior online and stressed the importance of cyber ethic [17]. Curbelo and Cruz [18], who analyzed topics for the development of information security and information assurance courses, argue that the ethics of offensive hacking cannot be taught as a single course and should be integrated throughout a curriculum. While information security curricula can be improved with more emphasis on cyber ethics, the case can be made for all university students to undergo basic information security training.

Another approach for introducing ethics, in addition to its inclusion in traditional information security programs through ethical hacking training, is the dissemination of information security concepts into other programs and coursework. White, et al. [13] discuss the process by which they changed their existing computer science curricula at the United States Air Force Academy by introducing various security related subjects into their traditional computer science courses. Their work promotes the notion that information security should be introduced throughout all introductory computer science courses with a particular focus on the "ethical and legal issues surrounding hacking and viruses" (White et. al, 1997). AlMalki and Al-Falayleh [19](2013) created an ethical hacking case study for teaching cyber using the systems and networks at the American University in the Emirates (AUE). They sniffed the AUE's network packets using a Man in the Middle attack (MitM) and were able to capture usernames and passwords of various AUE affiliated email accounts, POP3 email accounts, and AUE's Facebook credentials. They were also able to obtain some voice calls of the University's call center. This hacking was done for the purpose of creating the case study as well as exposing vulnerabilities. AlMalki and Al-Falayleh's work represents an example of ethical hacking content that can be incorporated into traditional defensive-orient information security courses for the purpose of a cyber ethics discussion.

The incorporation of ethical hacking into information security curricula does have its potential drawbacks, but in our opinion these are overshadowed by the gains in terms of the skill sets developed by students and the content of cyber ethics exposure they also receive. Many researchers agree that teaching hacking is a dangerous gamble and may involve many legal issues which may allow students to make incorrect decisions thereby creating liability for all parties involved (the student, the instructor, and the university). Separate work by Hartley, Livermore, and Durant [20–22] indicates that 90% of information security attacks happens from within an organization. This leads to the question of whether any "hacking" should be deemed acceptable within a university's network or systems. Wulf [23] indicates that "A problem with teaching undergraduate students using this (hacking) approach is that the instructor is effectively providing them with a loaded gun". According to Gross [24], "A hacker is anybody looking to manipulate technology to do something other than its original purpose. That's not necessarily a bad thing." There are numerous "grade changing" and similar cases at lower levels of the educational system involving easily implemented technologies, such as hardware or software keystroke loggers, where students took advantage of their technical knowledge in an unethical or criminal fashion. [25–28] As noted previously, although there is some risk involved with the introduction of ethical hacking to students, the benefits outweigh the risks. In addition, the troubling cases described above lend support to the notion that many students possess the technical skills to conduct attacks such as a PDOS or similar unethical online behavior in a university setting.

The Los Angeles Unified School District (LAUSD) recently initiated a $1 billion effort to provide every student in the LA area with an Apple iPad, thereby allowing students from low-income families to have equal access to such technology. [29] The devices were "locked down" so as to restrict access to certain websites and content and also were restricted to the school district's networks. Students were allowed to posses the devices outside of school, but the restrictions were intended to prevent device misuse and limit school district liability for such. In the Theodore Roosevelt High School, one of the first schools to participate in this program, 300 of the devices were hacked within the first week of program implementation.[30] Within this very short time period, students found a way to use their iPads for unrestricted personal use by deleting their personal profile information. Removing the personal profile information allowed the devices to be used on any network and access any website or information they chose. [30, 31] Stories such as these make the case again the teaching of ethical hacking in information security programs, but certainly make the case for cyber ethics and basic information security education for students at the university level.

4 Example of the PDOS Attack: A Need for Information Security Education for all Students

The case of a PDOS attack, as previously defined, represents an example of the type of cybercrime that can occur on a university campus. A cybercrime that is easily perpetrated and very hard to detect until it has been successfully carried out. Such attacks, which utilize online security mechanisms built into services that users access to "lock out" the rightful owner of an account, are easily perpetrated through a web browser, difficult to track due to IP address masking strategies, and represent an easy cyber crime to perpetrate on a university campus. While a simple PDOS attack can be conducted with nothing more than information that is readily available online, such as an email address, a more targeted PDOS attacks could be conducted when you integrate social engineering practices to acquire additional account information about a potential attack victim. Due to the fact that an attacker intentionally attempting to unsuccessfully login to a victim's online account appears to be normal Internet traffic on a university network, a PDOS, much like a lot of personal information security-related cybercrimes including phishing scams and the like, are not discovered until after it is too late for university network administrators to track properly and identify the attacker. In the book by the famous hacker and security expert Kevin Mitnick, "The Art of Deception: Controlling the Human Element of Security", the author discusses how people can be seen as the weakest link in security. [32] Even with limited knowledge of information security and the technical aspects of the Internet, a university campus environment would allow a student to gain enough information about what online services/websites another student uses in order to carry out an attack such as a PDOS. Gaining information on where an intended victim banks online or what online email service provider they use is as simple as shoulder surfing in a computer laboratory or dormitory room. It does not take a large amount of information, some of which is publicly available such as university email addresses, to conduct a PDOS. Social engineering and security awareness programs must become a common theme in university curricula for all students, regardless of their chosen field of study in order to address the ease with which attacks such as a PDOS can be conducted. Teaching only the technical side of information security in focused programs does not allow for the dissemination of information regarding personal information security and best practices for students when accessing online accounts.

In order to help ascertain the propensity of students to commit a cybercrime such as a PDOS, the authors developed a few survey questions related to account security breaches and attitudes of students towards them. These questions were included in a larger set of survey questions related to information security that were given to classes at four universities as part of a larger body of information security research by the authors. In order to have a "control group" with which to compare student answers to, the survey was also given to a group of law enforcement professionals attending a training seminar on cybercrime. The university students were all undergraduate students with two sets of them in computing/technical areas of (information science and information security) and two sets of them in business school programs. The goal of this limited survey was "proof of concept" in that the authors sought to support the notion that students do attempt to commit cybercrimes such as an account breaches (which might be considered a proxy for a PDOS), but in general believe that such crimes are not worth investigating if they occur. Some of the results are briefly discussed in an informal fashion below in order to bolster the argument for increased information security and cyber ethics training in universities for all students, not because the survey was the focus of this work.

As an example, a question in the survey was: "Have you ever attempted to login into another person's online account (email, online service, ecommerce website, etc.) without their permission?" In this question, and with the other questions of the survey, the account breach is a proxy for a PDOS. If a student attempts to log on to another's online account without permission, that same student may actually be committing a PDOS through a few unsuccessful attempts. Interestingly enough, 47% of the 115 total undergraduate students who answered this question said yes. In comparison, only 15% of the 60 law enforcement officials said yes. One might infer three possible reasons for the difference between the grouped sets of respondents. The first could be due to an age differential since the students were generally in their teens and early twenties in age while the ages ranged broadly from these same ages to fifties and sixties with the law enforcement officials. The differences in age might be incorporated in the next two reasons though. A second possible reason could be the potentially higher level of skill with respect to the Internet technologies and the ease with which such technology is used by the students when compared to the law enforcement officials. A third possible reason deals with attitudes and cyber ethics and legalities: that the students saw less of an ethical or legal problem with attempting to log onto someone else's account than the law enforcement officials. Regardless of which reason had the most impact on the question's results, it is clear that such an attempted cybercrime (account breach) is one that does occur with some regularity on university campuses.

A second question from the survey was directly related to the one previously described and asked: "If you answered yes to question one, was this attempt done in a joking fashion or as some sort of challenge to see if you could be successful?" Of the students who answered yes to the first question, 70% answered yes to this follow-up one. This is in comparison to 100% of the law enforcement officers. If one is to surmise that attempting an account breach as a joke or a challenge portends no malice towards the account's owner, then approximately 14 percent (30% of the 47% who answered yes to question one) of all surveyed students attempted at least one account breach with some form of malice as the intent. This result in particular provides support for the inclusion of information security and cyber ethics training for all university students.

A third question asked: "Are you aware of any laws relating to the process of attempting to use another person's online account?" This question is directed at discerning student awareness of the legalities of account breaches (as a proxy for the legalities of personal information security in general). Interestingly enough, 57% of the students answered yes to this question while only 47% of the law enforcement officers did. A greater awareness of the legal ramifications among the students only makes the result of question two even more troubling. The answer to this third question is further illuminated with the fourth survey question: "If you answered yes to question three, do you consider such laws deterrents (meaning that they would prevent you or other people from attempting such an activity because there would be a real possibility of being caught and prosecuted)?" In this case, less than half of the students responding (45%) thought that such laws were deterrents which coincides with the percentage of the law enforcement officials (46%). One can surmise from the results of these two questions that even though laws may exist in certain jurisdictions prohibiting the attempt of an online account breach, students do not take such laws as serious deterrents (and law enforcement does not as well).

Directly related to questions three and four, a fifth question asked: "If no malice is intended when attempting to log on to another person's online accounts, do you think it is a useful activity for law enforcement to investigate and pursue prosecution for such activities?" Again, the answers of the students were similar to those of law enforcement officials: 70% of students answered yes and 76% for law enforcement. This question addresses the sensitivity of students to an attempted breach on their own accounts. In other words, if a student believes that someone attempting to log onto their account without permission is a violation of their person information security, they would most likely answer yes to this question. The answers to this question are also interesting in light of the fact that both students and law enforcement do not believe existing laws to be deterrents. Despite this notion, they still want such breaches investigated and prosecuted if discovered.

A final question explores the propensity of university student to utilize a common tactic for obtaining account information to attempt an account breach. This question asked: "Have you ever employed a social engineering tactic to acquire someone else's account credentials?" The results of this question show that 16% of the students surveyed answered yes and only 7% of law enforcement did likewise. The 16% is similar to the 14% figure previously derived for the percentage of students attempting breaches with some form of malice. The results of this question make intuitive sense since one would assume that social engineering is the easiest tactic to deploy when attempting to obtain online account information from another student in a university environment.

While these questions were limited in scope and were part of a larger information security research survey, they do provide some evidence that cyber crimes such as a PDOS occur on university campuses. They also provide further evidence that students do not want to become victims of such crimes, as evidenced by the result in question five. These two pieces of evidence, when put together, provide support for the inclusion of basic information security in a personal context for university students. The survey questions also bring to light then need for instruction in cyber ethics. The fact that students do not think laws are deterrents creates the need for developing a sense of personal online responsibility and good practices for the university student body.

5 Summary

In the context of university campuses, where students unlimited access to Internet connectivity and are in close proximity with one another, there is great potential for cyber crime to occur. We examined the need for personal information security and cyber ethics training for all university students. This need is examined in light of the nature of certain types of cyber crime, such as a Personal Denial of Service attack, which are easily perpetrated and difficult to track or prosecute. In addition, the need for increased focus on cyber ethics in information security curricula on these same university campuses, in the form of ethical hacking training, is also addressed by the authors.

References

[1] F. Cohen, 'Computer Viruses', Doctoral Dissertation, University of Southern California, 1985.

[2] M. Zenko, 'Admiral Michael Mullen: Farewell and Thank You', Retrieved from http://globalpublicsquare.blogs.cnn.com/2011/09/29/admiral-michael-mullen-farewell-and-thank-you/ on May 29, 2014, 2011.

[3] White House, 'The Comprehensive National Cybersecurity Initiative', Retrieved from http://www.whitehouse.gov/issues/foreign-policy/cybersecurity/national-initiative on 29 May 2014, 2012.

[4] National Security Administration, 'National Centers of Academic Excellence', Retrieved from http://www.nsa.gov/ia/academic_outreach/nat_cae/index.shtml on 29 May 2014, 2013.

[5] M. Bartolacci, L. LeBlanc, A. Podhradsky, 'Personal Denial Of Service (PDOS) Attacks: A Discussion and Exploration of a New Category of Cyber Crime', *Journal of Digital Forensics, Security and Law*, In Print, 2014.

[6] L. Cohen, M. Felson, 'Social change and crime rate trends: A routine activity approach', American Sociological Review, 588–608, 1979.

[7] Privacy Rights Clearinghouse, 'Are You Being Stalked?', Retrieved from https://www.privacyrights.org/are-you-being-stalked on 29 May 2014, 2014.

[8] M. Schwartz, 'Cybersecurity Expert Shortage Puts U.S. A Risk' Retrieved from www.informationweek.com on May 15, 2014, 2010.

[9] National Security Administration, 'Criteria for Measurement for CAE / Cyber Operations Retrieved from www.nsa.gov/academia/nat_cae_cyber_ops/nat_cae_co_criteria.shtml on May 29, 2014, 2012.

[10] T. Cook, G. Conti, D. Raymond, U. Stated, M. Academy, 'When Good Ninjas Turn Bad: Preventing Your Students from Becoming the Threat', Proceedings of the 16th Colloquium for Information Systems Security Education, 61–67, 2012.

[11] S. Bratus, A. Shubina, M. Locasto, 'Teaching the Principles of the Hacker Curriculum to Undergraduates', 31st ACM Technical Symposium on Computer Science Education, ACM, doi:10.1145/1734263.1734303.

[12] D. Carnevale, 'Basic Training for Anti-Hackers: An intensive summer program drills students on cybersecurity skills', Chronicle of Higher Education, 2, 5, 2005.

[13] M. White, D. Ph, C. Gregory, L. Cohen, 'Security Across the Curriculum Using Computer Security to Teach Computer Science Principles' Internet Beseiged, ACM Press, 1997.

[14] K. Arnett, M. Schmidt, 'Busting the Ghost in the Machine', Communications of the ACM, 92–95, 2005.

[15] M. Dornseif, F. Gartner, T. Holtz, M. Mink, 'An Offensive Approach to Teaching Information Security: Aachen summer school applied IT security', Technical Report AIB 205.02.

[16] G. Vigna, 'Teaching Hands-on Network Security: Testbeds and Live Exercises', Journal of Information Warfare, 2(3), 8–24, 2003.

[17] Z. Trabelsi, W. Ibrahim, 'A Hands-on Approach for Teaching Denial of Service Attacks: A Case Study', Journal of Information Technology Education: Innovations in Practice, 12, 299–319, 2013.

[18] M. Curbelo, A. Cruz, A. 'Faculty Attitudes Toward Teaching Ethical Hacking to Computer and Information Systems Undergraduates Students', Eleventh LACCEI Latin American and Caribbean Conference for Engineering and Technology - Innovation in Engineering, Technology and Education for Competitiveness and Prosperity, 1–8, 2013

[19] M. AlMalki, M. Al-Falayleh, 'Ethical Hacking and Security Awareness: An Ounce of prevention is worth a pound of cure', Proceedings of Secure Abu Dhabi Conference, 2013.

[20] R. Hartley, 'Ethical Hacking: Teaching Students to Hack?', Doctoral Dissertation, East Carolina University, 2006.

[21] J. Livermore. 'What are Faculty Attitudes Toward Teaching Ethical Hacking and Penetration Testing?', Procedings of the 11th Colloquim for Information System Security Education, 2007.

[22] A. Durant, 'The Enemy Within'. Business *XL*, 2007.

[23] T. Wulf, 'Teaching ethics in undergraduate network', Consortium for Computing Sciences in College, 19(1), 2003.

[24] D. Gross, 'Mafiaboy" breaks silence, paints "portrait of a hacker', *CNN.com*, Retrieved on May 18, 2014, 2011.

[25] AssociatedPress, 'Monroe High School Students Caught Changing Grades' Retrieved from www.KOMONews.com on May 21, 2014, 2014.

[26] M. Birnbaum, J. Johnson, 'Students at Potomac school hack into computers; grades feared changed', The Washington Post, 2010.

[27] C. Gofford, 'Cheating scandal: Newport-Mesa official resigns to protest expulsions', LA Times, 2014.

[28] R. Wilkins, 'Grade-altering scheme sends ex-Purdue student to jail, 2nd student sentenced for hacking professors' computers', Retrieved from Jconline.com on May 2, 2014, 2014.

[29] H. Blume, S. Ceasar, 'Teachers union members, parents protest $1-billion iPad plan. Los Angeles Times, 2013.

[30] H. Blume, 'LAUSD halts home use of iPads for students after devices hacked', Los Angeles Times, 2013.

[31] A. Watters, 'Students Are "Hacking" Their School-Issued iPads: Good for Them. The limitations imposed on these devices inhibit students' natural curiosity', The Atlantic, 2013.

[32] K. Mitnick, W. Simon, The art of deception: Controlling the human element of security, John Wiley and Sons, 2001.

Biographies

Ashley Podhradsky is an Assistant Professor of Information Assurance and Forensics at Dakota State University in Madison, South Dakota. She received her D.Sc in Information Systems from Dakota State, with a specialization in information assurance and computer security. Her research interest include cyber security, specifically digital forensics. She has served as a guest editor for the Journal of Mobile Network Design and Innovation and the Journal of Interdisciplinary Telecommunications and Networking, both were special issues on Cyber Security. She is also the lead investigator at a security consulting firm in the Midwest.

Larry J. LeBlanc is a Professor of Operations Management in the Owen Graduate School of Management at Vanderbilt University. He received his Ph.D. from Northwestern University in Industrial Engineering/Management

Sciences. His research interests include analyzing spreadsheet risk, teaching management science using spreadsheets, supply chain analysis, spreadsheet optimization models, implementation of algorithms for large-scale optimization models, and telecommunication network design/analysis. He has 70 publications in referred journals and also 70 presentations at universities and organizations overseas. Dr. LeBlanc was an invited speaker at the INFORMS workshop on Teaching Management Science and has twice received the Dean's Award for Teaching Excellence. He was a guest editor of the special issue of the Interfaces on Spreadsheet Applications of Management Science and Operations Research.

Michael R. Bartolacci is an Associate Professor of Information Sciences and Technology at Penn State University - Berks. He holds a Ph.D. in Industrial Engineering and an MBA from Lehigh University. He conducts research in Information Security, Telecommunications Modeling, Information Technology Applications in Disaster Planning and Management, and Supply Chain Management.

www.ingramcontent.com/pod-product-compliance
Lightning Source LLC
LaVergne TN
LVHW012331060326
832902LV00011B/1834

9788793237216